WATER, LIFE, AND PROFIT

Water, Life, and Profit

Fluid Economies and Cultures of Niamey, Niger

Sara Beth Keough and Scott M. Youngstedt

NEW YORK • OXFORD
www.berghahnbooks.com

First published in 2019 by
Berghahn Books
www.berghahnbooks.com

© 2019 Sara Beth Keough and Scott M. Youngstedt

All rights reserved. Except for the quotation of short passages for the purposes of criticism and review, no part of this book may be reproduced in any form or by any means, electronic or mechanical, including photocopying, recording, or any information storage and retrieval system now known or to be invented, without written permission of the publisher.

Library of Congress Cataloging-in-Publication Data
Names: Keough, Sara Beth, 1976- author. | Youngstedt, Scott M., author.
Title: Water, life, and profit : fluid economies and cultures of Niamey, Niger / Sara Beth Keough and Scott M. Youngstedt.
Description: New York : Berghahn Books, 2019. | Includes bibliographical references and index.
Identifiers: LCCN 2019015308 (print) | LCCN 2019017986 (ebook) | ISBN 9781789203387 (ebook) | ISBN 9781789203370 (hardback : alk. paper)
Subjects: LCSH: Drinking water--Niger--Niamey. | Drinking water--Economic aspects--Niger--Niamey. | Drinking water--Social aspects--Niger--Niamey. | Bottled water industry--Economic aspects--Niger--Niamey. | Bottled water industry--Social aspects--Niger--Niamey.
Classification: LCC HD9349.M543 (ebook) | LCC HD9349.M543 N555 2019 (print) | DDC 333.9122096626--dc23
LC record available at https://lccn.loc.gov/2019015308

British Library Cataloguing in Publication Data
A catalogue record for this book is available from the British Library

ISBN 978-1-78920-337-0 hardback
ISBN 978-1-78920-338-7 ebook

Contents

List of Figures and Tables vi
Acknowledgments viii

Introduction
Why Water? Why Now? 1

Chapter 1
Situating Water in the Twenty-First Century 17

Chapter 2
Historical Urban Development in Niamey 40

Chapter 3
Accessing Water in Niamey 51

Chapter 4
Water Delivery Vendors in Niamey 71

Chapter 5
"Pure Water" in Niamey 96

Chapter 6
Fluid Materiality in Niamey 123

Conclusion 151

References 161
Index 175

Figures and Tables

Figures

0.1	Map of Niger within West Africa, with Niamey and major neighborhoods identified.	13
3.1	Locally made clay pots, called *tuluna*, for sale in a Niamey market.	52
3.2	Women at a well with installed pullies in Matankari.	54
3.3	Women and children collecting water from a standpipe serviced by a borehole in Matankari.	55
3.4	Children collecting water in *bidons* directly from a public standpipe in Niamey.	58
3.5	"Sodja Pompo" sign with standpipe in the background in Sonuci neighborhood, Niamey.	60
3.6	Sodja helping a customer fill *bidons* at his standpipe.	61
3.7	Kituba was Sunnah borehole and water tower across from Elhadji Mamane's compound.	65
4.1	*Ga'ruwa* at a public standpipe in Niamey.	73
4.2	A cluster of *ga'ruwa* carts around a public standpipe in a peripheral neighborhood of Niamey.	86
4.3	A *ga'ruwa* cart with decorations.	87
5.1	Mobile sachet water vendors, with a cooler containing bags on a pushcart in the foreground. The girl on the left has sachets in the bucket on her head.	107
5.2	A boutique with sachet water in bulk on the patio.	108
5.3	A *boutiquier* and his refrigerator filled with cold sachet water for sale.	109
5.4	Automated sachet-filling machine in a private home.	114
5.5	Close-up image of the automated process.	115
5.6	Sachet water in bulk (bags of twenty) stored in a garage waiting to be delivered to boutiques in Niamey and the surrounding villages.	116
6.1	*Ga'ruwa* cart with yellow plastic jugs.	126

6.2	Decorations on a *ga'ruwa* cart.	127
6.3	Plastic containers next to clay pots (from Hassane's compound).	128
6.4	Large neighborhood water tower (called a *chateau*).	130
6.5	Private water tower in a wealthy neighborhood compound.	131
6.6	Community private water tower in a poor neighborhood.	132
6.7	Advertisement for a borehole drilling company.	133
6.8	Lawn and sprinkler at the American International School of Niamey.	135
6.9	Lawn at the US ambassador's residence overlooking the Niger River.	136
6.10	Picture of Amico water sachet, commercially produced.	139
6.11	Billboard advertisement for Belvie bottled water.	141
6.12	Billboard advertisement for cups of water.	141
6.13	Billboard advertisement for water conservation (2015).	142
6.14	Billboard advertisement for water conservation (2016).	142

Tables

2.1	Population Growth of Niamey.	40
3.1	Cost of Water in Niamey by Transportation Method.	69

Acknowledgments

Water, Life, and Profit came to fruition through the efforts of many people who offered state, institutional, financial, temporal, emotional, and intellectual support. We are grateful to the République du Niger and its Ministère de l'Enseignement Superieur de la Recherche et de l'Innovation for offering us research clearance (N° 0747/MES/RI/SG/DGRI/DR). We thank Professor Abdou Bontianti, director of the Institut de Recherche en Sciences Humaines at Université Abdou Moumouni for being our institutional host and for assistance in navigating the complex process of securing research clearance. We also acknowledge institutional review board approval (IRB#: 534782-3) provided by our home institution, Saginaw Valley State University (SVSU).

We initiated our research on Niamey's water economies and cultures during month-long fieldwork visits in 2009–2010, 2011, 2012–2013, 2013–2014, and 2015. We conducted the bulk of the research during nine months of fieldwork in 2016–2017. Finally, we did follow-up work during a month-long visit in 2018. We deeply appreciate those who housed us and helped with logistics, especially Caroline and Kaocen Agalheir and Maggie Janes-Lucas.

Dr. Sara Beth Keough thanks the Fulbright Foundation, which made it possible to spend nine months in Niamey in 2016–2017. She also benefited from a research grant offered by the American Geographical Society. In addition, Dr. Keough greatly appreciates the Ruth and Ted Braun Foundation for providing her with a four-year research fellowship (2014–2018). SVSU provided Dr. Keough with professional development funds that supported language training and a sabbatical leave at the beginning of this project. Several of her early mentors, namely Ray Oldakowski, John Garrigus, Joe Scarpaci, Bon Richardson, Lydia Pulsipher, Tom Bell, and Peggy Gripshover, continue to provide support for her work, including this project, long after she ceased being their student. Scott M. Youngstedt appreciates the generosity of SVSU for providing a sabbatical leave in 2016–2017 and for three faculty research grants.

Our research and writing would not have been possible without the assistance and constructive input of many people. We are especially grateful for the assistance and guidance of three men in Niamey: Koche "Yaji" Dan Jima, Cheiffou Idrissa, and Alkassoum Alasmagui. Dan Jima served as an invaluable gatekeeper, which allowed us to find and interview many people critical to water economies in Niamey and in rural Niger that we could not otherwise have met. Social anthropologist Idrissa helped design our interview questions and also conducted interviews. Alasmagui's work was invaluable, as he conducted interviews with water vendors in Tamachek and Fulfulde—languages that we do not speak—and translated them into French. In addition, Djibrilla Garba offered useful advice and shared, in great detail, his own experiences in sachet water production. Abdou Aboubacar drove us to many neighborhoods to meet sachet water producers in their homes.

Dr. Hadiara Yaye Saidou, geography professor at Abdou Moumouni University, was a key supporter during Dr. Keough's Fulbright year, offering valuable advice and ideas for research direction. Ryan French and Ruth Dey, two students in SVSU's French language program, helped with transcription and translation. We also appreciate the helpful and regular feedback offered by Daniel Saftner and Eric Schmidt, whose time in Niger as graduate students with Fulbright grants overlapped with ours in 2016–2017. Perhaps most importantly, this book would not have been possible, of course, without the participation of hundreds of Nigériens who patiently answered our questions, permitted us to shadow them at work, and invited us into their homes.

Barbara M. Cooper, Hilary Hungerford, Baz LeCocq, Rachel R. Reynolds, and Joe Scarpaci provided enormously constructive advice at various stages of the writing process, and for this we express our sincere thanks to them. Members of Dr. Keough's Writing Accountability Group at SVSU provided immense amounts of support and encouragement, including several "shut-up-and-write" sessions during the final stages of this process. The detailed comments provided by anonymous Berghahn Books reviewers also served to strengthen the manuscript. Discussions with Kristín Loftsdóttir and Wendy Wilson Fall provided important insights on Fulani water cultures, and discussions with Souleymane Diallo and Susan Rasmussen contributed to our understanding of Tuareg water cultures. We also deeply appreciate the professionalism and support of Harry Eagles and Tom Bonnington, editors at Berghahn Books.

Last but not least, we are grateful to our parents, siblings, and children for their unwavering support. Wynde Kate Reese was an excellent listener and asked great questions that helped us think about our work in different ways. Jamila Youngstedt has grown up listening to stories about Niger and joined us there in 2016. Reid Youngstedt—our son—began accompanying us during fieldwork at eighteen months old in 2013 and consistently contributed patience, curiosity, and an unusual interest in our work for a (now) six-year-old. He still refers to Niger as "home."

Introduction
WHY WATER? WHY NOW?

In March 2012, the World Health Organization reported that the United Nations Development Program (UNDP) Millennium Development Goal (MDG) of halving the world's population that did not have access to clean, safe drinking water had been met ahead of schedule (WHO 2012). The UN Sustainable Development Goals (SDGs), also known as the Global Goals, which were designed to build on MDG successes, were instituted in January 2016 with funding and priorities set to continue through 2030. SDG Goal #6 specifically addressed the issue of global access to clean water and sanitation, and this goal is key to the success of other goals, like #2 (Zero Hunger), #3 (Good Health and Well-Being), #11 (Sustainable Cities and Communities), #12 (Responsible Production and Consumption), and #13 (Climate Action) (UNDP 2016a).

Despite efforts to address water access and quality issues on a global scale, hundreds of millions of people still lack this access, most of them in Africa south of the Sahara. MDG sanitation goals have not been achieved, and it has not been determined if the increased access proclaimed by the UNDP was achieved using methods that will be sustainable. Furthermore, the very definition of "access" is called into question, as it is often (and mistakenly) used interchangeably with "availability," and depending on the entity using the term, it can ignore issues of cost and time involved in accessing water that is available.

This book is situated within the broader global water development framework. Water access and quality problems are both broadly global and immediately local. As philosopher Thomas Pogge (2008) emphasizes, a combination of the global economic order and local conditions contribute to the current persistent poverty and water access problems that plague places like Africa south of the Sahara, and efforts to address these inequalities must consider how the two scales are related. Often, goals and programs developed by overarching agencies like the UNDP and nongovernmental organizations (NGOs) emphasize the global over the local. Our book seeks to do the opposite, by contributing to discussions of water access through a local ethnographic examination of water vending in one West African city. The cultural practices and livelihoods surrounding water production, access, and consumption are an essential consideration for the twenty-first century.

Water vending often fills the access gap in urban areas between those who can afford a connection to formal piped systems and those with the means to secure completely private access to water, such as through wells. Water-vending methods emerged from local perceptions about water; historically symbolic associations with water; gendered relations around water; the value and meaning of the materials involved in water production, vending, and consumption; local and global economic conditions affecting a local water regime; and cultural practices involving various aspects of urban life. Drawing from long-term ethnographic fieldwork by a geographer and an anthropologist, a woman and a man, conducted from 2013 to 2018, this book explores the tension between local cultural-historical ideas about water and its proper use, such as those expressed in the Qur'an (98 percent of Nigériens are Muslims) and the reality of water access in a neoliberal capitalist world.

Research on water quality, sanitation, health, and access is abundant, especially that which focuses on the world's impoverished populations. As global freshwater supplies become ever more compromised through pollution and overextraction, and as climate change redistributes atmospheric moisture and patterns of precipitation, studies on water access, sanitation, and health (WASH) move to the forefront of efforts to alleviate water-related problems. Conflict over diminishing water resources threatens several world regions, and the gap between those with access to clean, safe drinking water and those without widens further.

Water, Life, and Profit offers an original contribution to the water literature—much of which focuses on water quality and access in poor countries—through a holistic analysis of the people, economies, cultural symbolism, and material culture involved in the management, production, distribution, and consumption of drinking water in the urban context of Niamey, Niger. We draw from anthropology, geography, political economy, political ecology, and material culture studies. Although the examination is wide ranging, this book pays particular attention to two key groups of people operating in informal and hybridized economies who provide water to most of Niamey's residents: door-to-door water vendors (called ***ga'ruwa***) and those who sell water in one-half-liter plastic bags (sachets) on the street or in small shops. We explore the economics (management, production, distribution, and consumption) of each form of water delivery, focusing on the people involved and the symbolic meanings attached to the materials used in each stage of the process. Our analysis offers new insights on the lived experiences of gender, ethnicity, class, and spatial structure in Niamey's water economies today.

Although the focus of our research is local—two water-vending economies found in Niamey, Niger—it is at the same time global, as these economies are connected to and affected by international forces, flows, and structures. Here, water is the element connecting the social domains of those involved in its production, commoditization, and consumption (Orlove and Caton 2010), but it also connects the material, nonliving world with the living (Wagner 2013) and, thus, is essential to human activity and life (Strang 2013). In this way, water has power and is powerful (Hastrup and Hastrup 2016), and technology has enabled us to harness that power. In short, this book considers the social and lived nature of water as expressed through the processes of water vending and the lives of those involved.

While the connections to global forces offered by this book are numerous, the two water-vending economies discussed here are intricately linked to ideas of privatization, commoditization, and consumption to create conditions that affect the daily lives and profits of those involved in commoditized water production. Our research is based on the position that access to clean, safe drinking water is a fundamental human right, yet the reality of water provision in the twenty-first century is that accessing safe water is a commodity for which someone must pay, in part because the quality fresh water resources in the world are compromised and in part because the (now necessary) process of water purification and distribution requires labor and materials, thus commodifying this resource essential to human survival. Furthermore, neoliberal policies and practices have privatized, or created public-private partnerships, so that decisions about access to and distribution of water are made by corporate and government executives often driven by the for-profit and cost recovery realities of water provision. We now explore these interrelated realities of water, life, and profit from which the book title is derived.

Water

In April 2014, the city of Flint, Michigan, under the supervision of a governor-appointed emergency manager, switched the city's water supply from Detroit's treated municipal water to water from the Flint River. Within weeks, reported bacterial contamination led to water-boiling requirements for residents, and within six months, lead levels more than seven times the US Environmental Protection Agency's (EPA's) recommended limit were reported in water entering residents' homes through their taps. The magnitude of the crisis was soon realized when local pediatricians reported that the number of children with lead poisoning had doubled since the water supply switch, and in a few specific neighborhoods, the number of poisoned children tripled (Gupta et al. 2016).

Prior to this discovery, dozens of Flint residents reported water quality problems to officials, but their complaints were largely ignored. In the months following the January 2016 state of emergency declaration by then Michigan governor Rick Snyder, blame bounced from Flint's emergency manager to EPA officials to the governor himself (Davey and Smith 2016), and many speculated that the delayed reaction to the crisis was in part due to the fact that 41.9 percent of Flint's population lives below the poverty line and 54.3 percent are African American (US Bureau of the Census 2016). At the time of this writing, water in Flint is still deemed unsafe for drinking and cooking, as various investigations impede the implementation of concrete solutions for Flint's residents (Fonger 2018).

The water crisis in Flint, a city only forty miles from our university, raises several ethical questions regarding the human right to access safe drinking water: What water is safe to drink? How do we know? How, why, and where has safe drinking water become a luxury of the wealthy classes? What strategies do those without reliable access to safe drinking water use? What are the global and local conditions that led to these inequalities in wealth, health, materialities, and water provision?

In the Sahel in general, and in Niger in particular, fresh water is in short supply. Niger, a landlocked country that lies largely in the Sahara Desert, has very little surface fresh water. Only a narrow belt of land in the southern strip of the country receives much seasonal precipitation. Most Nigériens access water from subsurface deposits created long ago, when the Sahara region was more humid. In 2011, Niger ranked 111 (out of 180 countries) in renewable water resources per capita, including groundwater and surface water (Njoh and Akiwumi 2011). Yet only about 50 percent of Niger's water resources are renewable through annual precipitation, and this estimate is growing smaller as global climate change has shifted precipitation patterns in the Sahel (MacDonald et al. 2012).

Water is not just a physical substance with geographical patterns of distribution. It is a biocultural substance, something that connects the physical and human worlds. Veronica Strang (2015: 9) describes our engagement with water as a condition that is "as cultural as it is natural and, over time and space, the ways that societies have thought about, understood and acted upon water are in some ways fantastically diverse, and in others remarkably consistent." Both the compromised quality of freshwater resources and the power and policy structures in place that control its distribution have contributed to the commoditization of water.

The two water economies described in this book contribute to ethical discussions surrounding water access and the neoliberal realities that in part contextualize water access in the twenty-first century. Residents along the Niger River have access to a year-round supply of surface fresh water, and those living in Niamey have access to treated water, which is extracted from the Niger River and chemically purified before being pumped through the city's piped network. The chemical purification is an essential element in Niamey's water supply; however, because of water contamination in the Niger River, clean water in Niamey has become a commodity.

Life (and Death)

Availability of and access to safe drinking water is directly linked to issues of life expectancy, health, sanitation, and food security. Furthermore, by-products of water consumption create environmental hazards that further compromise the health of usually the poorest populations. Niger is no exception. Of the three leading causes of death for children under five years of age in Niger (malaria, respiratory infections, and diarrhea), two are directly attributable to water. Other health statistics in Niger, such as life expectancy at birth (sixty-one for females, sixty-three for males), are also linked to lack of access to treated water. It should be noted that these statistics show an improvement from their equivalents in the year 2000, as reported by the World Health Organization (WHO), but they still fall short of MDG and SDG targets (WHO 2015). The Institute for Health Metrics and Evaluation (IHME) reports that in Niger malarial and diarrheal diseases were the two leading causes of death in children under five years old between 2005 and 2016, and water access and sanitation risks are the leading factors in deaths and disabilities for the entire population during the same period (IHME 2018).

Water quality and availability are also related to issues of food security and soil quality; thus, they affect health and life in indirect ways as well. Periods of insufficient rainfall and overdrawing of ground water resources can compromise food availability and increase instances of malnutrition and famine and thus negatively affect health. Niger, a country situated in arid and semiarid climate zones, is particularly vulnerable to variations in rainfall. Genetically modified seeds are often introduced as a method to combat food insecurity, among other motivations, and this is certainly true in the Sahel. These non-native seeds often require more water than native seeds. The chemical fertilizers and pesticides needed to grow non-native species affect soil and groundwater quality.

Poverty is often an indicator for insufficient access to potable water, sanitation, and water for other uses such as bathing, cooking, and irrigation (Grant 2015). In 2016, 44.1 percent of Niger's population lived in poverty, and Niger ranked 187th out of 188 countries in per capita income, making it the second poorest country in the world (UNDP 2016b; World Bank 2017). In 2008, only 39 percent of Nigériens had "access to an improved source of drinking water in rural areas" (WSP 2011: 22)—where about 80 percent of the population resides. In 2008, "73 percent of the urban population had access to an improved source of drinking water," including 37 percent that had a household connection (WSP 2011: 26).

Even when sufficient and potable water is available, the cost may be beyond the means of a large portion of the population. Most of Niger's water supply is subterranean, and the costs for bringing this water to the surface are high. Niger also has one of the highest population growth rates in the world, 3.6 percent according to the World Bank, meaning that its water demands will only increase in future years. Moreover, according to some reports, Niamey is the fastest growing city in the world, "with Oxford Economics forecasting average annual population growth of 5.2% percent between 2015 and 2030" (*Guardian* 2015).

One by-product of the commoditization of water is packaging. While we explore the perceptions and symbolism involved in packaging water later in this book, it is worth mentioning here that packaging water creates further environmental hazards, hazards to which the poor are disproportionately exposed. For example, the plastic bags, or sachets as they are also known, used in the sale of cold water on the streets of Niamey are typically discarded in the area where they are consumed, usually on the side of the road, and we observed no attempt at reusing or recycling these bags during our fieldwork. The wind that whips through the Sahel blows these bags into ditches, where they prevent adequate drainage of sewage and water, leading to sanitation and health problems that affect human well-being. The plastic bags are also often burned, along with other trash and plastic products, creating localized toxic levels of air pollution.

Plastic pollution is mobile: "A staggering eight million metric tons [of plastic waste] wind up in oceans every year" and 93 percent of it comes from just ten rivers, one of which is the Niger River (Patel 2008: 1). We suspect that some discarded sachets travel all the way from Niamey to the Bight of Benin and beyond.

These examples show how poor water quality, access, and availability have negative consequences and how compromised access in Niamey has led to the emergence of water vendors. This book explores two types of water-vending economies wherein those involved earn livelihoods by producing and selling water in different forms. The income from water selling supports families, sends children to school, facilitates the participation in other elements of civil society, and gets reinvested back into the water-selling economy. For those involved in water economies, water becomes *financial* life.

Profit

The Second World Water Forum in The Hague in 2000 focused on water resource ownership and management and their impact on poverty, society, economic development, and the environment (World Water Forum 2000). Moreover, a key theoretical and practical debate about ownership and management of water has to do with financial models for water delivery systems and, with this, the question of whether water as a human right and profit objectives are compatible. The Ministerial Declaration that emerged from the forum solidified in its rhetoric the idea that water had become a commodity. The declaration stated goals of providing water "management" and access at "affordable prices" (Ministerial Declaration 2000). In short, it was decided that water could be sold for a profit because it was a human need, rather than a human right (Page 2005). This further opened the door for the private sector to profit from the distribution and accessibility of water.

Water is profitable in part due to its commoditization and because it is an essential life source, yet profits are realized in globally inequitable ways, which is why water commoditization is a topic of debate at all levels. As geographer Ben Page (2005: 293) explains, however, this commoditization "is not new, permanent, or inevitable."

Commoditization is, though, linked to water's materiality, or its physical attributes that "affect its relation to the human body and environment and that shape its use" (Orlove and Caton 2010: 403). Furthermore, the materiality of water is socially constructed. In other words, people assign meanings and values to water's materiality, and these meanings are incorporated into actions, such as water vending, that connect people in what Orlove and Caton (2010: 403) call "waterworlds." For example, in rural Niger, wells provide water to small villages and connect individuals who come to these central locations to get water. Although the water drawn from the wells is free, the wells themselves have a materiality because they shape the environment and human interaction associated with them. In contrast with rural Niger, water is a commodity in the city of Niamey because it is purified and distributed by both public and private entities to residents of the city. Residents must pay for water, whether they obtain it directly from the piped network or indirectly through the means we explain in this book.

Niamey's water regime is essentially controlled by four entities: two government agencies and two public-private companies. The Ministère de l'Hydraulique et de l'Environnement (Ministry of Hydrology and Environment) and the Ministère de l'Hydraulique et d'Assainissement (Ministry of Water and Sanitation) set policies and strategies for water distribution. The Ministry of Hydrology and Environment largely oversees water resources in rural areas, which is most of the country, although in some locations they work with community user associations that have some local control over their own systems. The Ministry of Water and Sanitation sets prices and policies in urban areas, including Niamey.

Falling under the jurisdiction of the two government ministries, Société de Patrimoine de l'Eau du Niger (SPEN) and Société d'Exploitation des Eaux du Niger (SEEN), are public-private partnerships that were created during water restructuring in 2000–2001 (Tidjani Alou 2005). SPEN is responsible for urban water infrastructure investment and debt service repayment, or the charges to customers SPEN includes in billing that directly help recoup upfront investments in infrastructure (Maiga 2016). It covers water distribution in fifty-two urban centers in Niger (Bardasi and Wodon 2008). SEEN operates under a lease contract to SPEN and works in partnership with the French water management company Veolia, one of the largest water management companies in the world. After water sector restructuring in 2000–2001, Veolia now holds a 51 percent share of SEEN (Tidjani Alou 2005; USAID 2010). SEEN operates public water services, including the purification and distribution of water to fifty-two urban areas in Niger, including Niamey, and is responsible for water service marketing (Maiga 2016; Veolia Inc. 2017). In short, the water supply in Niamey is delegated to SEEN by SPEN (Baron 2014).

As this book explains, the commoditization of water, and the profits realized through these processes, occur at multiple levels. Multinational corporations like Veolia control municipal water supplies in cities and regions all over the world, including Flint, Michigan. They work independently or in partnership with national and regional governments to implement distribution and marketing programs, recover costs, and realize profits. Some government agencies are truly public institutions and operate outside of public-private partnerships, but they are under pressure from entities at many levels to provide safe water without losing money. Formal distribution systems connect water resources with consumers either directly or indirectly through local individuals who fill gaps in access and availability within certain geographic contexts and with particular materials.

These small water enterprises (SWEs) (see Opryszko et al. 2009), such as the two water economies we describe in this book, are examples of local, individual entrepreneurs who are connected to global systems through markets, materials, and policies. The multiscalar complexities that are the reality of water access today create conditions of inequality, whereby the exploited and economically disadvantaged populations pay more for water than the wealthy (Bardasi and Wodon 2008). Referring to the US as well as global problem, Yanco (2014: 41) emphasizes that "it is expensive to be poor. It's not just that people living in poverty have less money to pay for basic necessities; basic necessities for the poor actually cost more." Thus, the water economies in Niamey provide a window into the lived realities of the poor in navigating water options, realities influenced by structures of access and inequality.

The Looming Water Crisis

The connections between water, life, and profit in Niamey occur within the context of a looming water crisis in the Sahel-Sahara and worldwide. Ironically, Niger sits on top of a subterranean water reserve that will be coveted across the Sahel region, but how to get access to it and at what cost remains to be seen. Cultural anthropologist John Bodley (2017: 325) identifies three crucial problems that threaten the global system: climate change and ecocide, inequality, and conflict. All three are contributors to the global water crisis. Furthermore, these problems are interlinked in many ways. For example, the poor bear a disproportionate burden of the problems associated with climate change, and this inequality leads to conflict over resources that will likely intensify in the future.

Climate Change

On the African continent, there is more variation in rainfall than in temperature, as most of the continent lies between the Tropic of Cancer and the Tropic of Capricorn. Climate change is anthropogenic and is negatively affecting the African continent, particularly through deforestation and desertification, pollution, and industrial emissions (Grant 2015). The Intergovernmental Panel on Climate Change (IPCC) predicts that although Africa as a region contributes little to global carbon emissions, its population and environment are one of the most vulnerable in the world, in part due to persistent poverty on the continent (Boko et al. 2007; Parnell and Walawege 2011; Williams 2015).

Temperatures on the continent are predicted to rise 1.5 times faster than the rest of the world. It is predicted that because of this rise the malaria zone will expand to above 2,000 meters in East Africa, the most densely populated region of the continent and an area that for now is malaria-free. Rising surface temperatures will affect farming and food security as evaporation rates from soil, rivers, lakes, and ponds increase. Rainfall distributions will shift, creating a necessary shift in types or species of crops produced (Williams 2015). In short, climate change will exacerbate the existing water crisis in Africa, affecting the economy and well-being of all.

Inequality

Industrialized countries are largely responsible for the majority of greenhouse gas emissions, while the world's impoverished populations pay the price. Current president of Uganda, Yoweri Museveni, calls climate change "an act of aggression by the developed world against the developing world" (quoted in Brown, Hammill, and McLeman 2007: 1142). As water resources are further compromised, especially in urban areas, these disadvantages play out in several ways. Travel and waiting times for water at public standpipes increase, a condition that disproportionately affects women. Even in areas with improved drinking water sources, the flow and availability of water may not be consistent across the day, week, or year. Finally, "improved" drinking water is not necessarily "safe" drinking water by definition, as transport or storage containers can still contaminate it and contain bacteria such as *E. coli*. Although the UN claims to have met the MDGs for water provision, the largest group of countries without access to improved drinking water is clustered in Africa south of the Sahara (Slaymaker and Bain 2017).

The scale at which data is reported can mask inequalities in water access. For example, the UN, UNICEF, the WHO, and others often report statistics at the country level (see Slaymaker and Bain 2017). This scale obscures geographic, gender, class, ethnic group, and neighborhood disparities in access. As a case in point, in Niamey the expansion of the piped water network often favors wealthier neighborhoods where for-profit water provision companies are more likely to recover their upfront investments in infrastructure. Thus, within Niamey, there are drastic inequalities in access by neighborhood and even neighborhood subunits.

Conflict

Conflict over scarce resources is not a new phenomenon and is an important element in the global water crisis. Heavy reliance on rain-fed agriculture in Africa means that changes in surface temperatures and precipitation could lead to increasing conflict over both water and food supplies. For example, changes in rainfall patterns have pushed some pastoralist groups onto land traditionally controlled by sedentary agricultural groups, leading to conflict, compromising their herds, and forcing them into sedentary, typically urban wage labor activities. Moreover, these changes have contributed to increasingly lethal confrontations between Tuareg and Fulani pastoralists, particularly in the Niger-Mali border region. In Niger, as we describe in Chapter 4, these conditions shape the lives of Tuareg and Fulani men who now work as water deliverers (among other jobs). Grant (2015) describes three categories of impacts of urban climate change: physical risks, such as sea-level rise and more frequent severe weather events; difficulties maintaining services, such as water, electricity, and sanitation; and the challenge of meeting the demands of increased urbanization, which is one of the effects of climate change.

In Niger, sea-level rise is not a concern, but severe weather events are. More intense storms and rainfall can diminish soil quality because runoff rates are often higher, particularly after a long dry season. Droughts and diminished or inconsistent rainfall can affect food security. And urban areas in Niger, including Niamey, are currently seeing exponential growth beyond the government's ability to provide services, particularly because Niger produces only a small amount of energy domestically and instead imports a large portion of its energy supply from Nigeria.

West Africa is a region with a high degree of water interdependence, as all countries except Cape Verde share at least one international watercourse (Niasse 2005). Dam projects, such as the Kandadji Dam on the Niger River in Niger, threaten water resources in downstream Nigeria, raising concerns and protests among the populations affected, including the government of Nigeria, which opposes any upstream dam project that reduces water flow in the river basin by more than 10 percent (Niasse 2005). Construction of the Kandadji Dam in Niger began in 2008 and is scheduled to be completed in 2020, and it continues to increase resource tensions between Nigeria and Niger.

Research Methods

The looming water crisis created by climate change, conflict, and inequality provides the context for our research on water vendors. Because issues of scale are essential to understanding the lived experiences of acquiring water, our neighborhood-scale study of water economies contributes hyperlocal perspectives on the relationships between water, life, and profit. We consider economies and livelihoods and their connections to water symbolism and concepts using local, on the ground experiences and realities in the context of national and global processes.

To consider the meaning-laden symbolism of water, the livelihoods that exist around its acquisition and dissemination, and the profits gained by its commoditization, we employed several research methods, most of them grounded in ethnography. This mixed methods approach allows a holistic interpretation of water economies, in which water is "integral, even essential, to many if not most domains or institutions of society—economic, political religious, leisure, etc." (Orlove and Caton 2010: 402; see also Björkman 2015: 15 and Strang 2004: 5).

Formal data collection for this book occurred over the course of several trips to Niger (Figure 0.1) between 2013 and 2018, although many of our ideas are also informed by prior field expeditions. We also lived in Niamey on a Fulbright Fellowship during the 2016–2017 academic year, and a majority of our interviews were conducted during this time. We both speak French, and Keough maintains intermediate skills in Hausa. Youngstedt, in particular, has been doing field research in Niger since 1988 and is nearly fluent in Hausa, which provides a longitudinal perspective on this topic, particularly on the material changes in water access over time.

We used a mixed methods approach in data collection with an emphasis on ethnographic methods, including literature review, participant observation, semistructured and structured interviews, shadowing,

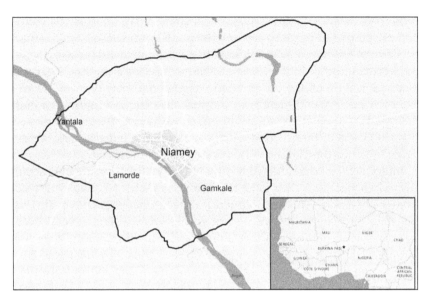

Figure 0.1 Map of Niger within West Africa, with Niamey and major neighborhoods identified. Cartography by Birch Bradford. Figure images are the intellectual property of Esri and are used herein with permission. © 2019 Esri and its licensors. All rights reserved.

archival research, and our own personal experience living in Niamey as a family with our preschool-aged son and as water consumers living in Niamey. Gatekeepers in several neighborhoods were essential assistants, as they were able to explain and validate our presence to those we hoped to interview. In the case of the water sachets, which are largely produced in private homes or compounds, gatekeepers were essential informants, as we could not tell from the street which houses had machines for making water sachets and which did not.

Thus, most neighborhoods included in this study are ones where we already knew several residents, where we were already established as researchers who could be trusted, and where we had gatekeepers who could help us make important connections. In ethnographic fieldwork, a random sample of neighborhoods is less important than the quality of data we can collect by using already established connections and rapport. A few neighborhoods were specifically excluded from our study, for example Koara Kano, the wealthy, largely expatriate neighborhood in Commune I, because water delivery and roadside water vendors do not exist there.

In total, we conducted 205 individual interviews and 8 focus group interviews involving approximately 50 additional people, with water producers, vendors, managers, and consumers. We interviewed some of these people and groups several times and spent weeks with them as participant observers. We explained to participants who we were, why we were interested in their role in Niamey's water economy, what we planned to do with the information we gathered, how we would protect their privacy and identities, and that they could withdraw or end their participation at any time without consequence. We gave explanations in Hausa or French, and when participants spoke neither of these languages, our translator communicated this information.

Interviews were conducted mostly in Hausa, occasionally in French, and when necessary, we paid local residents and graduate students from l'Université Abdou Moumouni (Niger's national university) to translate into languages we do not speak. We also had several field assistants who helped conduct interviews, allowing us to widen our sample and cover a larger geographic area. Thus, our data was collected from all five communes in Niamey (the main political division unit within the city, per the French urban development tradition), although Commune I, home to the wealthiest neighborhoods including a large embassy-affiliated expatriate community, is underrepresented compared to the others.

As we are interested in the material elements of these water economies, we also photographed means of water transportation, storage, vending, and consumption. Of particular interest to us were the mechanically produced water sachets (discussed in Chapter 4) and the labels on them. To this end, we collected sixty-five different brands that had been discarded on the side of the road after consumption. We also collected labels from eleven different brands of bottled water, including seven made in Niger, three in France, and one in Burkina Faso.

The neighborhood scale dominates our analysis for two important reasons. First, much less research on this scale of water economies than on larger scales exists. Comprehensive, citywide studies, such as that of Hungerford (2012) provide valuable insight that informs our study, especially as it provides information on household consumption, which was not a significant focus of our research. However, the neighborhood scale is appropriate for the study of hybridized forms of water vending and production because those forms are linked to assumptions about socioeconomic class. The neighborhood scale also reveals the lived experiences of inequality and how the complex relationship of ideologies of profit versus public good connects to cultures at various levels. In the postcolonial era in Niamey, and many other West African cities, neighborhood division is more strongly associated with socioeconomic class than race or ethnicity (Grant 2015).

Second, despite many attempts, we were able to secure only two short interviews with representatives of agencies involved in water distribution in Niamey: Société de Patrimoine de l'Eau du Niger (SPEN), Société d'Exploitation des Eaux du Niger (SEEN), Ministère de l'Hydraulique et de l'Environnement (Ministry of Hydrology and Environment), and Ministère de l'Hydraulique et d'Assainissement (Ministry of Water and Purification/Sanitation). Although secondary data and other published materials from and about these agencies and ministries provided the institutional and policy-specific background we needed, our primary data lacked a firsthand institutional perspective. Again, policy-oriented water research is plentiful, so our focus on water vending at the neighborhood level adds a less common perspective to the literature.

Overview of the Chapters

Our exploration of water, life, and profit through the lens of hybrid water economies in Niamey begins in Chapter 1 with an explanation of the thematic frameworks that inform our research. Specifically, we explain key ideas related to water governance, water access, and the symbolic nature of water that inform and influence the water economies we describe in detail later in the book. These three themes are, of course, interrelated, and there is significant overlap between them. Trends in governance affect access, or the lack thereof, the latter opening opportunities for alternative forms of water access. Each individually and collectively affects water, life, and profit in Niamey.

Chapter 2 situates Niger in general, and Niamey in particular, historically and geographically in a larger West African and developing world context. Interwoven into the history of Niamey's urban development are specific references to the evolution of the city's water regime, including the impact and influence of colonial powers, postindependence authoritarian governments, the establishment of democracy, and the implementation of structural adjustment programs and neoliberal practices.

In Chapter 3, we describe the different ways residents of Niamey access water. We emphasize that accessing water in Niamey involves a cross section of methods and strategies and varies by the purpose for which water will be used (drinking vs. washing clothes). We also examine the pricing structure of piped water in Niamey, a city that uses increasing block tariffs, to demonstrate why the poor end up paying more for water than the wealthy, even though they use less.

The next two chapters, Chapter 4 and Chapter 5, provide in-depth analyses of the two hybrid water economies that are the focus of this book. Both emerged to fill gaps in water access created by the governance patterns we describe in Chapter 1. In Chapter 4, we explore the realities of water, life, and profit among the door-to-door water vendors in Niamey, called *ga'ruwa* in Hausa. This method operates very differently in Niamey than in other African cities, and in addition to explaining these differences, we also consider why this job is dominated by Tuareg and Fulani men. Furthermore, we consider how ideas about gender, ethnicity, and Islam are integral in understanding this particular water economy.

The second hybrid water economy we analyze is the production of sachet water, the one-half-liter plastic bags of water that are mechanically filled and sealed and sold cold on the streets of Niamey. This is the focus of Chapter 5. In this chapter, we work backward through the commodity and value chain created and established through the production, vending, consumption, and discarding of plastic sachets as well as the groups involved in each part of the chain. At each part of the commodity and value chain, we consider the symbolic nature of water and this form of plastic packaging, particularly as it relates to livelihoods and profit.

Although we touch on materiality in Chapters 4 and 5, we focus intensely on the material culture of water in Chapter 6. In addition to exploring the symbolism of materials used in the two water economies highlighted in Chapters 4 and 5, we explore the branding of water. To this end, we analyze four highway billboards in Niamey: two commercial and two public service messages. We also analyze the labeling used on water sachets, and we connect these efforts at branding to broader links between packaging and ideas of purity and confidence.

Finally, the conclusion offers some final thoughts about water, life, and profit as they pertain to trends in governance, access, and materiality. We offer a critique of Niamey's water regime, suggestions for pro-poor water policies, and warnings that failure to consider the sociocultural elements surrounding water when attempting to improve access could create more problems than are solved.

1

SITUATING WATER IN THE TWENTY-FIRST CENTURY

Shortly after a state of emergency was declared in Flint, Michigan—due to lead and other contaminants in the water supply—in January 2016 by then governor Rick Snyder, the state government (and later the Flint city government) began distributing bottled water for drinking and cooking free of charge to Flint residents. (Residents had to pay for the tap water used for bathing, washing clothes, and other purposes.) Between January 2016 and the program's end in April 2018, the state spent more than $350 million to provide bottled water for drinking and cooking to Flint residents affected by the water quality problems that emerged in 2014. This amount was in addition to $100 million in federal funding and includes spending on pipe replacement, healthcare, nutritional food distribution, education programming, and job training (Carmody 2018; Oosting and LeBlanc 2018).

By September 2017, the State of Michigan had provided more than 157 million bottles of water to Flint residents (Glenzain 2017), and that number continued to grow until the state ended the program in April 2018. Most of the bottled water was extracted and packaged by Nestlé, the largest beverage company in the world. Just two hours away from the city of Flint, Nestlé pumps more than 100,000 times the amount of water an average Michigan resident uses in a day from an underground aquifer and bottles it under several labels owned by the company, including Ice Mountain and Pure Life (Glenzain 2017). Nestlé pays just $200 a year for these water rights, and in April 2018 their plan to increase pumping from 250 gallons per minute to 400 gallons per minute was approved by the Michigan Department of Environmental Quality (MDEQ). More than 80,000 public comments against the plan were submitted and categorized by theme. The top three concerns expressed involved corporate greed versus people and the environment, that water should not be sold for profit, and the trend toward water privatization (Chappal 2018).

Nestlé's approval from the state came at a time when the costs of piped water to consumers in Flint were at an all-time high and thousands of residents in Flint and Detroit (27,000 homes in 2016 alone) had their water supply cut off by the city for delinquency in bill payment (Glenzain 2017). In addition to high costs for piped water and the looming withdraw of state-provided bottled water, the plastic pollution from these programs has extensive negative effects on the environment, the extent to which has not yet been realized. In 2016, for example, CNN reporters tallied 157 bottles of water used in a single Flint household in one day (Zdanowicz 2016). Even though the State of Michigan declared the water crisis over in Flint during April 2018, evidenced by hundreds of water quality tests that indicate Flint water falls below EPA standards for heavy metals and other contaminants, residents of Flint do not trust government officials, their reports, or the water itself (Glenzain 2017).

The water crisis in Flint exemplifies the complexity of global water quality and access issues. It speaks to practices of governance, policy making, financing, and security at local, state, and national levels. The Flint example reveals the variations in power relationships between public entities, elected officials, state agencies, private for-profit multinational corporations (who benefited substantially from the Flint water crisis and the length of time it took the state to repair the problem), utility companies and their subsidiaries, and community residents. It demonstrates how humans and the environment are inextricably linked, especially in relation to natural resources needed for survival. Ultimately, this case illuminates cultural dimensions of perception of the materials involved in the water crisis and the packaging and branding of "solutions" such as bottled water provision.

This chapter explores how three thematic frameworks inform the ideas about water, life, and profit described in the introduction. These thematic frameworks include urban water governance (including public, private, and community levels); access to and quality of water; and the cultural symbolism and material culture of water. Our goal is to illustrate in subsequent chapters how these three themes are implicated in stories of water, life, and profit in Niamey.

Urban Water Governance

In what is arguably the newest (as of 2018) regional geography text on Africa, author Richard Grant (2015) takes a new approach to studying the continent, one whereby urban areas are the primary focus of inquiry, rather than states, countries, and regions. He rationalizes this shift in African geographies by proposing that cities offer a better understanding of national and regional trends, and global connections, than other scales because of the "urban revolution's" impact (Pieterse and Parnell 2014; Myers 2011).

This focus on the societal implications of intense urbanization also extends the discussion to the connections between rural and urban areas; rural areas do not exist in isolation from urban ones—they are intricately connected economically, through kinship ties, through expanded transportation networks, and through new forms of communication technology. Finally, Grant (2015) argues that a more intense focus on cities as the scale of analysis in Africa further challenges the ways scholars have approached Africa as "abnormal." In this sense, much older work on Africa viewed the continent as atypical because it is a product of colonialism, capitalism created worse conditions in rural areas than urban ones, and pro-urban policy and spending would negatively affect investment in rural areas (and their natural resources).

Within this new focus on urban Africa, issues of governance are of particular concern, especially the issues related to water. Water is both life and power, and governance of water resources is highly contested among users. In the twenty-first century, governance of urban water supplies in Africa often involves the integration of public-private partnerships that emerged out of several decades of structural adjustment programs and market liberalization, and that established hybrid economies of water vending and small water enterprises and informal means of access, such as water sharing. Central to discussions of urban water governance is the integrated water resource management (IWRM) paradigm, which promotes "the coordinated development and management of water, land and related resources in order to maximize economic and social welfare in an equitable manner without compromising the sustainability of vital ecosystems and the environment" (Global Water Partnership 2018).

Although criticized for being yet another Western development model implemented with varying results in African cities and overlooking local cultural and political dimensions, IWRM is informed by the idea that the "crisis" is not in water itself but in management of water resources (Munck 2015). It attempts to link ideas of availability, sustainability, potability, local cultural norms, and politics with water management at various scales. In short, IWRM is one element of the umbrella discourse on water that calls for good governance as the solution to water poverty and inequality (Bakker et al. 2008).

Transnational organizations are often responsible for promoting paradigms like IWRM. The World Bank, for example, assumes that private sector participation (PSP) is the key to alleviating water poverty (Bakker 2007), although there is strong opposition to this idea, a theme discussed later in this section. In essence, though, the World Bank has evolved from being an institution that, in the 1950s and 1960s, primarily granted small loans to governments in the Global South to pay contractors from the North to build infrastructure related to development schemes, into an entity that today has adopted poverty alleviation as its primary focus, albeit within the boundaries of development capitalism and profit promotion. In the 1950s and 1960s, however, the practice of granting small government loans introduced Northern investors to the Global South (then called the Third World) as a source of "untapped and unvalorized natural resources that could potentially fuel a tremendous growth spurt of Northern industrial output and profit" (Goldman 2007: 788).

Loans came with strings attached, the most notorious being those that fell under the structural adjustment umbrella of the 1980s to the present. Water privatization became a key focus beginning in the 1990s, and the World Bank's global water policy of today consists of three main practices: creation of transnational policy networks, development loans that encourage the privatization of water and sanitation services, and imposing conditions upon borrowers who need access to capital (Goldman 2007). In short, the World Bank (and its member countries) became a key player in global water governance.

As evidenced by World Bank activities, water governance is more than just a technical issue or matter of provision—it is a social and political issue revolving around power disparities between advantaged and disadvantaged populations. Yet the developmentalist paradigm sees water as a resource to be harnessed and, especially in the 1990s, as an economic good to be exploited. Wittfogel's (1957) "hydraulic despotism" speaks to elite control over water resources, and Tony Allan (2003) criticizes the "hydraulic mission," the dominant discourse from the late nineteenth century until at least 1980, for portraying water as a technical issue that led to unsustainable development of irrigation systems through dam construction and marginalization of community opposition (antidam) groups as the answer to food security concerns (Munck 2015). Large overarching global efforts, like the UN's International Decade for Clean Water and Sanitation (shortened to the Water Decade), Millennium Development Goals (MDGs), Sustainable Development Goals (SDGs), and now the "New Water Decade" (2018–2028) are often criticized for overlooking these social and political contexts and for failing to consider needs of individuals and groups in different regions and settings (Munck 2015).

The Evolution of Water Governance

There are several distinct periods of water governance programs implemented by the development community in postcolonial African countries (and other parts of the Global South). Munck (2015) argues that most of these programs ignore the impacts of colonial era water provision, which privilege elite expatriate neighborhoods and the colonial-controlled city center over indigenous ones, which were often peripheral. This dual segregate model of urban form in French-speaking Africa became more complex in the 1960s and 1970s after independence. Water access during this time was linked to land tenure status and governed by local public authorities and community organizations (Baron 2008).

The decade between 1980 and 1990, also labeled the "Water Decade" (and now the "First Water Decade" to distinguish it from the "New Water Decade" of 2018–2028) by the UN, promised increased awareness of and support for global efforts to provide clean water and sanitation. Yet this was also the decade that began structural adjustment, setting the stage for (or becoming the driving force behind) the privatization trends of the 1990s. Water governance increased in complexity during this decade, as international institutions like the World Bank and the International Monetary Fund (IMF) became involved. Privatization was the trend, and urban areas were treated as homogenous regions of official settlements, ignoring residents who lived in city sectors without land ownership titles (Baron 2008). The extent to which the programs of the 1980s and 1990s were successful is debated, as the programs that began in these decades deemed water an economic good with its price regulated far less by governments and more by the free market like other commodities (Munck 2015).

A central paradox of these water decades is the tension between ideas about water as a public good and water as a private commodity. And, of course, the rhetorical work of the UN and transnational organizations on water is part of this paradox, especially in the twenty-first century. The 2000s marked a partial withdrawal of the private sector from urban water provision, as anticipated profits had not been realized, especially from large impoverished areas of African cities. Public-private partnerships (PPP) became the more common model of urban water governance, although Baron (2008) argues that many practices of the 1990s continued into the new millennium. Regardless, the overall failure of privatization to reduce disparities in access (Munck 2015) inspired the UN's MDGs, which guided international development up through 2015.

These goals, many of them unmet, were then reformulated as the UN SDGs, which will theoretically guide global development through 2030. The SDGs, two of which relate specifically to water and many of which indirectly rely on access to clean water for their success, overlap with the UN's "New Water Decade" (2018–2028). Efforts to meet the goals have called into question the sustainability of "completed" projects; the extent to which projected population counts (natural increase and migration) are considered; and the extent to which, as Munck (2015) argues, social and political contexts of local areas and groups are incorporated. Furthermore, neoliberal water governance models have not transformed urban water regimes or coverage rates on the continent (Bontianti et al. 2014; Sambu 2016). In short, the poor continue to be the victims of failed Western initiatives (Munck 2015).

Governance Failure

Governance, and governance failure, is a key area of focus within neoliberal water models (Bakker et al. 2008; Munck 2015), yet there is debate over the role of the public and private sectors in water supply management. The push toward privatization relied on the assumption that PSP would lead to increased access to water and sanitation facilities because the private sector had the means to invest in infrastructure upfront, which would improve performance and efficiency. This efficiency would, in turn, motivate and encourage individual investment in connections, especially among the poor. In reality, though, PSP has meant reduced spending on repair and replacement of infrastructure as well as increasing costs to consumers in an effort to realize profits (Munck 2015). The public sector does not present a good governance record either, especially when the state is both the supplier and the regulator, but even when the regulator is private. Corruption plagues both the public and private sectors.

These conditions are what Bakker et al. (2008: 1894–95) describe as two related things: "market failure" and "state failure." Failure to significantly improve connection rates in poor urban households in the Global South during the late 1990s and 2000s led to debates over the regulations and laws governing water use and distribution and the ownership of water resources, including the merits of private sector, public sector, and community-based ownership. Traditionally, research on water governance focused on ownership, but 21st-century scholarship has prioritized the role of institutional factors, or "the political, organizational, and administrative processes through which stakeholders (including citizens and interest groups) articulate their interests, exercise their legal rights, take decisions, meet their obligations, and mediate their differences" (Bakker et al. 2008: 1894). According to Bakker, governance failure occurs when institutions fail to support pro-poor water supply policies. There is, however, disagreement about where responsibility for failed water provision and regulation falls. Supply failure, for example, occurs in several interrelated arenas, including states, markets, and governance in general (Bakker et al. 2008; Bakker 2010).

What Is Good Governance?

By the early 2000s, many corporations realized that recovering costs in water provision was not as lucrative as they first anticipated, and they began to reduce (though not abandon) their role in water provision or move toward PPPs with state institutions. Although neoliberal discourse continued, it now competed with stronger voices arguing for governance standards that promoted access to clean water as a human right and basic need, better informed pro-poor policies, sustainable management practices, and stronger contributions from African governments and experts. On Tony Blair's Commission for Africa, for example, a majority of commissioners were African leaders (Commission for Africa 2016).

Neoliberal discourse also competed with consideration for policy impacts on women and sector-wide coordination in water governance, such as IWRM, while at the same time being mindful of market trends and opportunities for profit (Munck 2015). Infrastructure, including water networks, states Björkman (2015), cannot be viewed simply in its technical purpose but must be designed to facilitate access to what Simone (2004: 407–8) calls "spaces of economic and cultural production" and a means for living life in the city. This "soft path" approach considered water "not as an end-product, but as a means to accomplish certain tasks" (Wade 2012: 215). These practices have become the standard by which "good governance" of water are now measured (Munck 2015).

Promoters of "good governance" are concerned by the influence of the World Bank. Thus, an emphasis on incorporating indigenous knowledge of and experience with water resources and working within existing local (African) political frameworks rather than imposing a top-down, North to South approach to water governance became part of the "soft path" approach to governance. Ken Conca (2006), in his book *Governing Water*, promotes the idea of an institutional structure and decision-making process that limits negative impacts of decisions because the structure can accommodate a variety of interests (and interest groups), while at the same time treating water as a shared resource.

While the long-term impact of the IWRM paradigm, generally considered a "soft path approach," remains to be seen (Gleick 2003; see also Mtsi and Nicol 2015 for tentative outcomes in East and Southern Africa), scholars like Swatuk (2015) and Munck (2015: 14) believe a strong developmental democratic state must be a key player, perhaps the primary player, in water provision. This is because they are "the only entity capable of addressing the needs of the poor for sustainable access to safe water." Furthermore, they propose a water governance policy that distinguishes between "practical water needs," which refer to the minimum amount of water required for basic community needs, and "strategic water needs," which include social needs and typically emphasize decommoditization of water, with different management plans for each.

Niger has been subject to many of these trends in water governance. As described in the introduction to this book, water governance is both centralized and decentralized, as power is shared between two government ministries: a private company owned in part by a French multinational corporation and PPPs. NGOs and humanitarian aid organizations often shoulder the responsibility for improved water access in rural areas, but in Niger, they are largely absent from urban water provisioning landscapes.

Informal or hybrid forms of water access, such as the ones we discuss in this book, fall somewhat outside (but not completely outside) regulatory frameworks designed by government and private entities. This creates a fractured urban water regime that, in general, favors wealthy neighborhoods over the practical and strategic water needs of poor neighborhoods. It is these informal and hybrid forms of water access, however, that most embody the objectives of IWRM, as they improve access and efficiency, especially for women. These hybrid forms take into account local cultural values and perceptions involving water, increase agency as they offer several choices for Niamey residents to access water, and involve local, neighborhood-level forms of management and decision making. However, water governance is not the only factor in discussions of water access. The next section highlights ideas of water access and quality that influence the realities in Niamey.

Water Access and Quality

In the preface to *Social Power and the Urbanization of Water,* author Erik Swyngedouw (2004) recounts his first visit to Guayaquil, Ecuador in 1992. Standing in one of Guayaquil's informal settlements, where he had a panoramic view of the city, he describes the irony of the city's hydraulic landscape: the Guayas River flows through the city, carrying millions of gallons of water each day, yet for hundreds of thousands of people, accessing safe water is a daily struggle. It is a struggle to simply acquire water in neighborhoods that lacked even public standpipes, and it is a struggle to afford the exploitive prices of water charged by vendors and a struggle to consume enough clean water to stay healthy.

The provision of safe, potable water across the globe is considered a major development challenge (and in some cases, a visible failure) thus far in the twenty-first century (Munck 2015). Furthermore, water provision does not just refer to availability; it includes issues of access, including physical, environmental, and financial factors. Discussions about water provision, access, and quality are directly linked to ideas about governance, especially in urban areas, where most of the available water for personal use is controlled by public or private entities, and occasionally by community groups. Furthermore, it is often the inequalities of water access that open the door for informal or hybrid forms of access, such as the water-vending economies we discuss in Chapters 4 and 5.

It makes sense to begin a discussion of water access by considering infrastructure. The previous section explored issues of governance (local, national, global) that affect water supplies and policies in urban Africa. This section is less about top-down control of water regulations and more about the opportunities for, and realities of, water access in urban Africa, particularly for the poor. This discussion will then provide a framework for understanding the two water economies in Niamey that we discuss in the remainder of this book.

Current water infrastructure in most African cities has colonial roots. Njoh and Akiwumi (2011) found that among major African cities, the length of the colonial period was a partial indicator of the degree of access to improved water and sanitation facilities. Their study showed that the longer the period of colonization, the greater the access. This was, in part, due to the desire by Europeans to re-create in African cities the conditions they enjoyed in European ones. Location of informal and unplanned settlements during the colonial era, as well as the nature of city expansion, also influenced patterns of access (Dill and

Crow 2014). In the period after independence, the idea of the state as utility provider continued until the 1980s, when principles of neoliberalism and structural adjustment programs changed governance strategies of utility provision. This was followed by a blending of public and private enterprises in the 1990s and a fracturing of services in the 2000s whereby informal and hybrid providers filled gaps in access (Baron and Bonnaissieux 2011).

Although piped networks designed and constructed by Europeans varied substantially across the continent (for a comparison between French and British systems, see Hungerford and Smiley 2016 and Njoh 2008), the general trend in provision during the colonial era involved what Graham and Marvin (2001) refer to as "splintering urbanism," wherein infrastructure projects and their locations were highly selective and biased toward the colonizer before the colonized. Policies regarding water services and costs reflected the social inequalities that emerged from this stratified system.

In general, a pattern emerged in colonial African cities in which primarily public standpipes existed in African neighborhoods and direct connections to homes dominated European areas (Baron and Bonnaissieux 2011; Hungerford and Smiley 2016). This selective infrastructure planning led to what Karen Bakker (2003, 2010) describes as "elite archipelagos" of access, whereby pockets or islands of connection to piped water exist in a sea of areas without access. She argues that this term is more accurate than "networks," because "networks" imply a continuous line of connection along the infrastructure in place. Furthermore, she challenges the relevance of Graham and Marvin's "splintering urbanism" because it is grounded in the assumption of widespread water service provision, which is not the reality in many African cities. As our research in Niamey shows, access to water is not a linear phenomenon, as spatial exclusion or marginalization exists across the water spectrum (Smiley and Koti 2010).

While colonial infrastructure created spaces of exclusion in water access and neoliberal practices have introduced a variety of players in urban water regimes, Catherine Baron and Mahaman Tidjani Alou (2011) argue that these trends have increased the importance of local approaches to water access. These local approaches, including sachet water and water delivery approaches that are the focus of this book, are more commonly found in poor neighborhoods in African cities (Baron 2014), although as we show with sachet water, they are not exclusive to those areas.

The idea of connectivity is part of the story of water access. Water itself is a connective substance because it is present in virtually all elements of society and can facilitate interaction between people—a "total social fact" (Orlove and Caton 2010: 402; see also Mauss 1950 and Strang 2004). In other words, as people work to obtain water, they encounter several domains of society—bureaucratic and political, social, environmental, infrastructural, legal—with which they have no choice but to interact. Thus, changes in, for example, governance or infrastructure, not only affect access to water, they also affect other societal domains and social interaction. In short, a piped network is not the only infrastructure of water provision in an urban area; people also serve as infrastructure (Simone 2004) through their social interactions as some work to provide water to those without reliable access to water's physical infrastructure.

Amartya Sen's (1999) and Martha Nussbaum's (2008) "capability approach" is relevant to this discussion, as it helps to explain why different individuals are able to access and harness the power of certain commodities (like water) to achieve particular goals (Robeyns 2005). It emphasizes individual agency and personal and cultural values and preferences (Ibrahim 2006). In other words, while there may be external factors that limit an individuals' capability to access a water supply network, there is also the possibility that an individual may choose, for personal and/or cultural reasons, not to connect to a network that is otherwise available (Bakker et al. 2008). In many African cities, numerous obstacles to accessing potable water exist for the poor. These obstacles often are physical and geographic (distance from potable water and time required to obtain basic amounts of water for daily living) as well as institutional and structural (extent of piped network and costs of taking water from it).

Regardless of how urban residents obtain water, we must consider the social interactions involved in these processes to understand the larger issues of water access and quality. The relational-dialectical approach to water production employed by Linton and Budds (2014) asserts the importance of the relationship between water and society. The "hydrosocial cycle," a derivation of the hydrologic cycle, seeks to reconnect the societal and physical scientific elements of water and processes of acquisition. These cycles are not bounded by physical infrastructure because they include, for example, exchange relationships, customer demands and preferences, cultural perceptions of water quality from various sources, laws that govern quality and pricing, ideas about water held by various belief systems, and rainfall patterns and climate change (Bakker 2003).

These hydrosocial cycles have created what Boelens et al. (2016) refer to as "hydrosocial territories" at various scales. Hydrosocial territories have water as a focal point but include and are bounded by the interactions between people in the act of obtaining water; the regulations and entities governing access and availability (or the absence of these regulatory bodies); the institutions controlling finance, infrastructure, and distribution of water resources; and the spaces created by the interaction of these forces. In other words, water and society are coproduced (Jepson et al. 2017) simultaneously at several scales.

Because society and water are inextricably linked, questions of water security must be considered multiscalar as well. For example, water rights that are sold by states to multinational corporations affect water supplies for residents in the immediate vicinity of the water source and change the social interactions that result from and are facilitated by the process of acquiring water. In urban areas of Africa, water acquisition also occurs at multiple scales, as poor households typically acquire water from several sources, including direct connections (when affordable), public standpipes, water deliverers, neighbors, boreholes, wells, and packaged water. This diversification of water access points implies a water regime that is less secure, at least for disadvantaged populations, because there is not one main reliable source of water for those whose incomes fluctuate. However, at the same time, this diversification perhaps helps ensure at least some access to water because if one source becomes unavailable (such as water to public standpipes, which may not flow all day or all year long), other sources of water exist to fill the gap.

Several household- or neighborhood-level water provision modalities have emerged in African cities as a response to situations of water insecurity and vulnerability among urban populations, as well as some of the governance and provision problems described earlier in this chapter. These alternative water provision modalities usually function alongside formal water infrastructure, such as a piped water and sewer network, and are often dependent on or connected to these formal systems. It then becomes difficult to distinguish between formal and informal systems of water provision, as forms often considered "informal" may rely on piped water and/or an electrical grid in order to function (Schwartz et al. 2015). Bakker (2010) argues that no truly "informal" systems of water provision exist, and she promotes the term "hybrid" as an appropriate label for the alternative forms of water delivery that characterized disadvantaged neighborhoods in cities of the Global South. These hybrid forms of water provision compete, in some urban areas, with natural water sources, such as rivers, that provide a free source of water and with formal systems of water delivery, leaving the hybrid forms in an even more precarious position (Baron and Tidjani Alou 2011).

In general, however, as we explained in the introduction to this book, the poor pay more for water than the wealthy because the poor are forced to buy water from alternative providers that lack the large economies of scale enjoyed by the state and corporations (Bardasi and Wodon 2008). We use our ethnographic studies of household water delivery and sachet water vending in Niamey to illustrate these trends in Chapter 4 and 5.

In some cities, the gap in water access is somewhat reduced, though not eliminated, by pro-poor water policies. Examples of pro-poor policies include prepaid meters (for an example in Kampala, Uganda, see Berg and Mugisha 2010), expansion of free public standpipes in informal settlements, successful monitoring by a supporting institution or organization, ensuring consistent water pressure during peak usage times, and subsidized costs for connecting to the piped network. Nickson and Franceys (2003: 111) propose that if the subsidies that normally go to wealthy customers were redirected to the poor as an incentive to increase the number of new connections, with a "rebalanced cost-reflective tariff structure that allows for recovery of capital costs, then the sector would generate the necessary funds to cope with population growth."

In Niamey, there are few incentives for utilities to serve the urban poor, but certain kinds of assistance are available, including subsidized costs to households for direct connections to the piped network, social tariffs, and equalization systems such as increasing block tariffs (see Nickson and Franceys 2003: 114; Baron 2014). However, public standpipes in Niamey are not free, and management of them has been delegated to neighborhood organizations who select individuals to serve as standpipe managers. This model is problematic, as managers may have little accountability to consumers and can exercise considerable control over standpipes and their use (Keener et al. 2010). We discuss a specific situation in which this occurred in Niamey in Chapter 4.

Pro-poor water efforts often exist in tension with the neoliberal practices of private multinational corporations (MNCs) (Laurie 2007). Several myths surround pro-poor policies in the water sector. One myth is that the Global South should be the focus of pro-poor policies (Laurie 2007), but examples from developed countries such as the United States prove that developing pro-poor water policies should have a global focus (Wescoat et al. 2007).

Another such myth is the idea that PSP in water provision, particularly in the form of the World Bank, the IMF, and MNCs, will increase efficiency of service and lower costs; yet research has shown that this pattern has not emerged (Nickson and Franceys 2003; Keener et al. 2010; Goldman 2007; Keough and Youngstedt 2018; Youngstedt et al. 2016). In reality, the involvement of these entities has not delivered on the promises made (Bakker 2007; Goldman 2007; Bardasi and Wodon 2008; Sambu 2016). In Jakarta, for example, new connections were most often in middle-class households and neighborhoods, and tariff increases were higher (Bakker 2007). In fact, when private entities failed to realize profits from urban water regimes, some changed the language in their contracts from, to cite an example in Bolivia, "piped connection" to just "connection," whereby the latter could refer to proximity to a public standpipe rather than a direct household connection (Goldman 2007: 796).

Governments sometimes use similar strategies. For example, the Oman government census measures the percentage of people with access to clean water, not the percentage of people with piped water in their homes (Limbert 2010: 117). This is a useful distinction even if done for political reasons. Hall and Lobina (2007) argue that a well-organized, democratic public sector is less vulnerable to market fluctuations than private enterprises and thus has a better chance of ensuring water provision to the urban poor, and Goldman (2007) highlights the World Bank's position that states should regulate, but not run, public services like water provision.

Kariuki et al. (2014) provide evidence from Kampala that when undergoing institutional reforms first, and improving utility performance second, pro-poor urban policies are more likely to be sustainable in the long term. However, many of these overarching water policies, such as those imposed by the World Bank, assume continuity across urban spaces. As Baron (2014) points out, the assumption is often that peripheral neighborhoods are in the most precarious position in terms of water access because piped networks have not expanded to these newest neighborhoods of African cities, when in reality, inner-city neighborhoods can be just as precarious if household incomes are not consistent or reliable. As the previous section of this chapter demonstrates, there is not a clean separation between the public and private sectors. The reality is that the water sector in African cities is complex, fragmented, and incomplete.

Symbolism and Materiality of Water

Issues of symbolism and materiality associated with water further complicate the condition of water resources in African cities. While the governance, accessibility, and availability of water is well documented, we aim in this book to make a contribution to the less-explored discourse on the symbolic and material nature of water and the relevance of semiotics to the study of water relationships, specifically within the context of Niamey. The economic value of water is often assigned based on the costs associated with acquisition, purification, and distribution—in other words, economic value is often assigned through the process of commoditization and, in the neoliberal era, privatization (Page 2005).

But water holds cultural, symbolic, and material value as well. These values can be understood by considering its social construction, or a holistic consideration for all the social domains, including value and meaning, in which water is used in society (Orlove and Caton 2010: 402–3). This "materiality" of water can exist alongside, in opposition to, or in light of its market-determined value. Water links humans and their physical environment, such as in the hydrosocial cycle discussed earlier in which water has connective qualities and through material symbolism because "water is simultaneously an economic input, an aesthetic reference, a religious symbol, a public service, a private good, a cornerstone of public health, and a biophysical necessity for humans and ecosystems alike" (Bakker 2010: 3).

In the introduction to his edited volume *Materiality*, anthropologist Daniel Miller (2005: 4) asks, "Can we have a theory of things?" As a professor of material culture studies, it is not surprising that Miller's answer to his own question is "yes," but he goes on to argue that things we are less aware of more powerfully control our behavior and environment without us seriously questioning their role. This power that things have contributes to their value, and in the case of commodities, the politics involved in their exchange and valuation constitutes a "social life" of things (Appadurai 1986). Anthropologist John R. Wagner (2013) concentrates specifically on water as a commodity whose social life crosses human and ecological domains, results in connections between people, and has the power to influence politics, policies, and economics. Thus, water has material agency, value, and meaning and can be considered an element of material culture because it is something we can purchase or possess (Berger 2014).

How water is acquired, how it is assigned both economic and symbolic value, what materials become associated with it, and what we can learn about humanity by studying how it is used (Buchli 2002) further solidifies its inclusion in discussions of material culture. Material culture is often studied through the lens of semiotics, and the remainder of this section explores the symbolic nature of water in general and water in Niger in particular.

Water is an archetypal symbol. It is essential to life but also has properties that "inspire metaphorical and poetic thought" (Hanchett et al. 2014: 2) and meanings that, in many cases, have remained consistent over time (Strang 2004). Part of this consistency is drawn from the fact that water exists in nature and can be used in its raw form. Although we usually purify water today for drinking, we can use water directly from a source for bathing, washing, and cooking without changing its chemistry. This idea that nature has semiotic applications is explored by anthropologist Mary Douglas (1970), who argues that nature in general offers especially potent resources for symbolism and that the cosmos and the body are always symbols of society. Water, perhaps to a greater extent than any other natural symbol of the cosmos, is "good to think with" (Lévi-Strauss 1966).

Water also serves as a "key symbol" (Ortner 1973: 1339)—a product of culture that also shapes culture. Key symbols are distinguished from less powerful symbols by virtue of the fact that the people say that they are, and they appear in a variety of contexts, conversations, and different symbolic domains (Ortner 1973: 1339). In *The Meaning of Water*, Strang (2004: 49) argues that people are inspired by the remarkable properties of water:

> The most constant "quality" of water is that it is not constant, but is characterised by transmutability and sensitivity to changes in the environment. Physically, it is the ultimate "fluid," ... shrinking and disappearing in the earth or evaporating into the ether. It has an extraordinary ability to metamorphose rapidly into substances with oppositional qualities, that is, the highly visible, concrete solidity of ice, and the fleeting dematerialisation of steam. Each state is endlessly reversible, so that this polymorphic range is always potentially present. In every aspect, water moves between oppositional extremes: it may be a roaring flood, or a still pool, invisible and transparent, or reflective and impenetrable. It may be life-giving, providing warm amniotic support and essential hydration, or it may burn, freeze or drown.

Creation of meaning often involves sensory perception, among other things, and people often find water visually, aurally, and tactilely mesmerizing. In fact, Strang (2004) argues:

> Of all the elements in the environment, it is the most suited to convey meaning in every aspect of human life... . There is a discernable consonance between the characteristics of water, people's sensory experience of these, and the conceptual schemes in which water appears as powerful metaphor.

Several key themes recur in water symbolism and cultures among peoples across time and space. Some examples include water gods and water worship in ancient and contemporary mythology, the central role of water origin stories, and water rituals, such as making sacrifices and offerings to rivers and the use of prayer and magic to bring rain (Strang 2004: 84). The duality of water—that is, its capacity to bring life and death—is also widely recognized. Most communities recognize the analogous importance of maintaining water balance in human bodies as well as ecosystems (Strang 2015: 33). In *Water Culture in South Asia: Bangladesh Perspectives*, one of relatively few ethnographic studies focused on water symbolism and culture, Hanchett et al. (2014) document specific Bangladeshi examples of the quasiuniversal themes identified by Strang (2004, 2015). Bangladeshis worship a wide range of water deities that are typically referred to with feminine metaphors. Much of their mythology involves the power and mystery of water. Rituals involving water are common, including rainmaking rites. Bangladeshi languages are infused with water words, stories, sayings, and songs about water. Furthermore, Bangladeshis associate water with health and illness and value holy water.

Water also plays a role in expressing social hierarchy (as well as leveling it). In the Hindu caste system in Bangladesh, for example, those in higher castes may give water to those in lower castes, but they will never take water from lower castes (Hanchett et al. 2014). Furthermore, while some Bangladeshi Muslims are regarded by Bangladeshi Hindus as constituting a very low caste, Bangladeshi Muslims do not recognize a caste system among themselves. Among Bangladeshi Muslim villagers, "*giving* water to anyone who is thirsty is promoted ... as a great virtue. This is regarded as a life-saving gesture and an obligation of a good person, one that will reward the giver in heaven" (Hanchett et al. 2014: 48). "Anyone" includes non-Muslims, enemies, and people of any social class.

Nigérien water symbolism and cultures parallel those of Bangladeshis. What makes the homologous consistency particularly fascinating is the vast disparity in rain between the two countries: Niger's defining problem is drought, while Bangladesh's defining problem is flooding. Most Nigériens and Bangladeshis are Muslim, and we will discuss the importance of water in Islam at the conclusion of this section.

People living along the Niger River—the source of Niamey's municipal water supply—have their own different names for it and their own myths about it. These typically emphasize both its physical and supernatural power. In Niger, the Songhay and Zarma have lived along the Niger River the longest. For them, the Niger is a life-giving river as well as a mystical world of danger, mystery, power, and spirits. For example, traditional Songhay tell stories of "Harakoy Dikko, the spirit queen of the Niger River and mother of the Tooru, nobles of the Songhay spirit world" (Stoller and Olkes 1987: 234). Some Songhay believe that powerful sorcerers can spend days under the Niger River in spirit villages where they learn secrets of power and healing (Stoller and Olkes 1987: 163–65).

Hausa have made up one-half of Niamey's population for about three decades due to migration, even though Niamey is outside the traditional boundaries of Hausaland. Some Hausa communities have lived for centuries adjacent to the Niger River in Northern Nigeria, where they recount stories of powerful river spirits. One of their most famous stories describes how access to water is the foundation of Hausa civilization—though it does not involve the Niger River. The Bayajida legend recounts the travels of Bayajida of Baghdad, who first traveled west, then south across the Sahara to the town of Daura located in contemporary Nigeria (Hallam 1966). At the time he arrived, a snake lived in the town's well and often denied the people access to the water and terrorized the population. Bayajida courageously slew the serpent. As a reward, the local queen, Magajiya Daurama, married him. Their offspring, together with Bayajida's children from other wives, went on to establish the Hausa Bakwai, or the original seven Hausa states. (As traditional nature religions faded and Christianity and Islam with their humanized deity spread, "a spate of serpent slaying [stories] followed" [Strang 20015: 84], and the Bayajida legend is representative of this corpus. These stories typically involve male culture heroes defeating the chaos of nature.)

In another example, Mami Wata (pidgin English for "Mother Water")—a water goddess typically thought to appear like either a mermaid or a serpent and recognized in many parts of Africa and the African diaspora—also resides in the Niger River (Drewal 2008). Masquelier (2008) examines one variant of Mami Wata in Niger that left her watery home to reside far from water in the Sahel, complete with hooves to facilitate her terrestrial adaptation. Local traditional religions of Niger feature water rituals and rituals involving water's cleansing and healing powers. Spirit possession music typically includes the use of calabashes floating in water—contained in either larger calabashes or plastic buckets—as percussion instruments due to their unique resonance. Some sorcerers are believed to have the power to control rain.

Nigérien languages include many different words for water, depending upon what holds it and moves it among other variables. The globally ubiquitous expression, "Water is life," is found in Nigérien languages—*ruwa ne rayuwar* in Hausa and *hari katifundi* in Zarma—and many proverbs use water symbolism metaphorically. In Hausa, the most widely spoken language in Niamey, the word *ruwa* (water) can be used as a synonym for important personal business. Indeed, one of the most emphatically delivered phrases in the Hausa language is *ba ruwanka!* (literally, "It is none of your water!"; figuratively, "It is none of your business!"). It is used to rebuke people who are regarded as nosy or rude. During evenings in Ramadan, Hausa commonly use the greeting *"barka da shan ruwa"* (Greetings on drinking water), suggesting that water is more difficult to forgo than food—a biological human truth. Finally, across Niger, water is a symbol of hospitality. It is always the first thing that hosts offer to guests.

Water also often functions as a "sacred symbol." "Sacred symbols function to synthesize a people's ethos—the tone, character, and quality of their life, its moral and aesthetic style and mood—and their world view—the picture they have of the way things in sheer actuality are, their most comprehensive ideas of order" (Geertz 1973: 89). The concepts of holy and healing waters are also remarkably widespread. As a "sacred substance," water has been and is used in "baptisms, libations, holy ablutions, fertility rites, for blessing and protection from the 'evil eye' and for mortuary rituals…. There is a similarly consistent relationship between the 'vitality' imputed to water and its capacity for healing" (Strang 2004: 85, 96).

In a predominantly Muslim country like Niger, water serves as an important sacred symbol in Islam. "The Qur'an states that God made water the basis of creation" (Strang 2015: 40) and "according to the Qur'an, every living thing is made from water" (Caponera 2001), endowing water with a sacred quality since Muslims believe that Allah created all living things. Islam emphasizes the "human responsibility to properly care for this gift from God. Sharing water is a good deed, and not sharing is a sin" (Faruqui et al. 2001, cited in Hanchett et al. 2014: 36). The crucial role of water in Islamic spiritual life is revealed in *shari'a*, the term generally translated as "Islamic law" or the "right way" also technically means "access to a source of pure water" (see Varisco 1983: 369–70; Limbert 2010: 123). In another example, it is believed that the Prophet Muhammad said people hold three things in common: water, pasture, and fire, and "The man who holds back water from another will have God's mercy held back from him" (see Varisco 1983: 369–70; Limbert 2010: 123).

Nigérien Muslims follow the global Islamic practice of performing ablutions before each of the five daily prayers. Drawing from Allah's instructions, this ritual involves the washing of the face, head, arms, and feet with water—a substance believed to purify Muslims in preparation for prayer. Only water has the power of purification for prayer, though the Qur'an allows the use of clean sand for ablutions in the absence of access to water. Furthermore, as Hanchett et al. (2014: 37) point out, "Head-to-foot washing with water is required of men and women wishing to perform the rites of pilgrimages to Mecca and other holy sites." In addition, Muslims ritually wash bodies with warm water before burial.

In Niamey, most men perform ablutions in small groups in public places outside of mosques or on street corners. Women typically perform ablutions individually or with female family members in the privacy of their homes. Many men regularly pray at three to five different places daily since they often move about the city due to work, meeting with their favorite conversation groups, and commuting. Men away from home at prayer time can count on men with whom they are working or visiting (or complete strangers) to provide them with enough water for ablutions. The religious sharing of water promotes a remarkable sense of solidarity and community among Nigériens that crosses ethnic and class lines. If boys are around, they are expected to bring water to adults. The vast majority of men and women in Niamey use small plastic teakettles imported from China to perform their ritual cleansing. Others use recycled tomato paste cans, typically imported from Nigeria and Côte d'Ivoire. Since water and its purifying qualities are conceived as a gift from Allah, water is used efficiently and sparingly. Nigériens we observed typically use only 240 to 360 ml of water for each round of ablutions.

Muslims are taught that only Allah bestows rain. In many recent years, Niger has held national Islamic days of prayer for rain. For example, in 2006:

> Following a call from Niger's Islamic council, religious leaders, or Imams, led worshippers—including President Mamadou Tandja—in special prayers conducted in the open air rather than in mosques. And mystical religious leaders, or Marabouts, encouraged worshipers to do away with their usual prayer mats and press their hands and foreheads directly on the dry parched earth for greater prayer success. (ReliefWeb 2006)

Similarly in 2011, President Mahamadou Issoufou took part in a "national collective prayer … asking for rain in a televised ceremony at Niamey's grand mosque led by Sheikh Djabir Ismael, president of the

AIN, Niger's largest Islamic association.... Prayer sessions were held across the country" (Modern Ghana 2011).

Special types of water have always been revered in Islam; for example, "Early links between water and eternal life are also evident in the ancient Islamic story of Al-Kidhr's discovery of the Well of Life, which reappears in many myths as the Fountain of Youth" (Strang 2015: 78). Sacred wells named after Muslim prophets containing water with miraculous healing properties are found in many places. The most famous of these wells—mentioned in a Hadith—is the Zamzam well located just twenty meters from the Ka'ba in Mecca from which Muslim pilgrims bring holy water home "in hopes that it will help to fulfill wishes, cure illness, succeed in childbirth, and otherwise help solve life's problems" (Hanchett 2014: 37).

Saudi Arabia prohibits the export of Zamzam water for sale, but the Kingdom does allow pilgrims to bring small quantities of it home (including precollected Zamzam water packaged in plastic bottles), and trucks transport some of it in tankers to Medina. More broadly, Islam opposes the commoditization of the commons. The sacredness of water is expressed through Islamic prohibitions on its sale; however,

> this is not to say that one cannot own water; the general rule is that water belongs to the person who first exploits it, to whomever undertakes the enterprise of digging or carrying it. However, one can only use what one needs and cannot make a profit from selling the rest.... Speculating is illegal, and portering water is acceptable only if one does not profit from it. (Limbert 2010: 123)

In today's capitalist world, many Muslims have commoditized water or live in countries where water is commoditized and privatized. Entrepreneurs in many countries sell bottled water using the Zamzam name, including some that make explicit claims that they are selling authentic holy water from the Zamzam well. This includes Mahavir Wholesale, a company that sells 500 ml bottles of "ZamZam" water for $9.99 each, which they claim is bottled in Saudi Arabia and distributed from Florida (Zifiti 2018). We found two brands of Zam-Zam sachet water in Niamey. Zam-Zam is a brand that is made in the Dogondoutchi region, according to its label. It features an advertising label in Hausa, *Tsab-tac-cin Ruwa Sha* (Clean Drinking Water).

Niger Lait SA in Niamey, a large company that has, since 1994, long specialized in producing and selling milk and yogurt drinks in plastic bags, manufactures the other brand of Zam-Zam. Their bags feature the slogan, *L'EAU PURE ET NATURELLE C'EST LA VIE* (Pure and natural water is life), a color image of a waterfall at a desert oasis, and the claim that their water has been *Traitement bactericide par UV et Filtration* (Treated with ultraviolet bactericide and filtration). Niger Lait also sells Zam-Zam in 250-ml plastic cups with an aluminum foil lid (similar to those used by airlines; see Chapter 6).

Over time, water has been commoditized in Nigérien cities, and paying someone to deliver water has become acceptable to urban Nigériens. It is now regarded as a basic social service. Most of the participants in our project were unaware of Islamic prohibitions against selling water. Some who were aware of Islamic prohibitions rationalized this situation by saying that they are not paying for water itself, but rather the delivery service (pipes or porters), just as Omanis did (Limbert 2010: 125). However, our interlocutors consistently indicated that Muslims should never have to pay for water at a standpipe or elsewhere to perform their ablutions before their Islamic prayers.

To return to our original discussion of water's symbolic nature, then, the religious symbolism attributed and assigned to water carries different meanings for different groups and can vary by context, yet there has been consistency over time. The prominence of water in Nigérien folklore and forms of Islam practiced in Niger guide actions related to water and in some ways stand in contrast to the commoditized and privatized conditions under which water is distributed. Thus, individuals attempt to rationalize water's materiality and commoditization against or within elements of Nigérien culture that support its decommoditization. Understanding how water is governed, accessed, and symbolized helps us understand the livelihoods surrounding water, life, and profit in Niamey.

Having now established trends related to water governance, access, materiality, and symbolism, we transition to our case studies of water economies in Niamey. After providing an overview of historical urban development in Niamey in Chapter 2, including the evolution of the city's water regime, we explore the different methods residents of Niamey use to access water in Chapter 3 and the general economic structures of place, drawing from our personal experience and fieldwork. Next, we explore the system of door-to-door water delivery in Niamey in Chapter 4, followed by the sachet water economy as it plays out in Niamey in Chapter 5. And, finally, we consider specific materialities of water that transcend these different modes of access, including the impact of plastic packaging and how packaged water is branded and marketed. These case studies highlight the complex relationships between water, life, and profit and are situated within the 21st-century frameworks of water governance in African cities, issues of access and commoditization, and ideas surrounding water's symbolic importance in Nigérien culture. In other words, themes of water governance, access, and symbolism run through all of these stories of water, life, and profit in one African city.

2

Historical Urban Development in Niamey

Niamey has grown exponentially, from a few hundred people in 1900 to about two million people in 2018 (See Table 2.1). France chose Niamey as the colonial capital of Niger, and it continued to serve as the capital after Niger gained its independence in 1960. Over this time, it became very ethnically diverse due to its political status and particularly as a result of migration from all parts of Niger. That is, much of its growth and diversity has been due to migration. In fact, through the last 120 years of its history, first- and second-generation migrants have constituted a substantial majority of the population. Most migrants up until about 1930 came from western Niger. Subsequently, Niamey gradually became a magnet for Nigériens living throughout the country. Many migrants came looking for work and the bright lights and adventure of living in the city, but the massive exodus of rural populations in the wake of drought and famine were also central factors in Niamey's rapid growth (Gado 1997: 35). These included famines in 1901–1903, 1913–1915, 1931–1932, 1954, 1969, 1972–1974, 1984–1985, and 2005 — all of which are named in local languages. Thus, Niamey has long served as a place to begin a new life as well as a refuge, a last resort for survival.

Rural to urban migration is, of course, not unique to Niger. The capitalist world economy locates capital and commodities in cities embedded in nation-states — even on the periphery of the world system — leading to the globalization of urbanization. By the late 1980s, migration accounted for 73.2 percent of Niamey's rate of population

Table 2.1 Population Growth of Niamey.

1900	1926	1960	1988	1997	2005	2010	2018*
600	3,142	33,816	400,000	700,000	859,000	1,222,066	2,000,000

Source: Gado 1997; République du Niger 2010; The Guardian 2015. *Authors' estimate.

growth (Cohen 1989: 69). Migration, in turn, influenced new residential patterns. Waves of migrants from across Niger settled wherever they could find affordable lodging in Niamey; so by about 1980 (and ever since) the city has been thoroughly residentially integrated by ethnicity. However, the city is segregated by class and thus by access to the city's piped water network—a topic that we will address in future chapters.

Hausa migrants played a central role in the rapid growth of Niamey from a small city of 33,816—of whom only 3,600, or 12 percent, were Hausa in 1960 (Sidikou 1980; Bernus 1969: 35, Decalo 1989: 165)—to a bustling West African capital of two million residents today. Hausa became a slight majority by about 1980 and have maintained that position up to the present. This constitutes a highly unusual transformation, given that Niamey rests some 150 kilometers west of the westernmost boundaries of traditional Hausaland.

Niamey and Its Water in Historical Context

Diverse peoples have shaped Niamey's unique identity and character through time. For hundreds of years (and probably more) until the turn of the twentieth century, Niamey consisted of a collection of small fishing villages—Goudel, Saga, Lamordé, Bitinkidji, and Gamkalé—situated alongside the banks of the Niger River and inhabited by a few hundred Zarma, Songhay, Mawri, and Fulani peoples (Gado 1997: 9, 14). The countryside surrounding Niamey was (and remains) primarily Zarma country, with some interspersion of Fulani and Tuareg. The approximately 250-kilometer stretch of the Niger River that flows through southwestern Niger is the country's only permanent river. It became the source of Niamey's municipal water system in the 1920s–1930s, as it flows through the city and has been a key conduit of fishing, irrigation, commerce, empire building, and cultural exchange in West Africa. From its headwaters in the Guinea highlands it flows north into the Sahel and Sahara in Mali, where it takes a southerly turn (the Niger Bend) near Timbuktu that takes it through Niger, Benin, and Nigeria before it enters the Bight of Benin through the famous Niger Delta.

Arguably, the most remarkable part of this 4,180-kilometer-long river is the roughly 1,000-kilometer stretch of it that winds its way through the Sahel in Niger and Mali (and even into the Sahara in Mali). In a place that receives only about 300–500 millimeters of rain annually in a three- to four-month rainy season from roughly June to September, this mighty river that never ceases its flow appears to defy the laws of physical geography.

Several stories recount the founding of the settlement and the origin of the name "Niamey" (Gado 1997: 14). For example, a Mawri story explains that after leaving Neini, their island home on the Niger River, the original Mawri settlers first rested and then settled on the river bank near a huge tree known as a *nia*. The tree had magical qualities, and though long since gone, its site on the current location of Hotel Gawaye is still remembered. In one Zarma version of the story that emphasizes domesticity, Zarma people established Niamey at the *nya-me* (mother's riverbank), where an older woman regularly collected water. Idrissa (2009: 162) argues, "The name Niamey ... derives from a Zarma-Songhay word meaning (what else?) 'intermingling.' This name has become over the decades a self-fulfilling prophecy, as other populations taken into the colony and Republic of Niger sent in waves of settlers to the capital."

Niamey was not included, at least by name, in any of the accounts of nineteenth-century European exploration (Decalo 1989: 171), including those of Barth, Monteil, and Toutée (Gado 1997: 13). It did not occupy the attention of the French empire until 1898, when the Hourst military reconnaissance mission first recorded the name "Niamey" (Gado 1997: 23). French forces took control of Niamey in 1901 and established a military supply post; named Niamey "Capital City of the Zarma Administrative Zone" (Chef-Lieu de Cercle du Djerma) in 1903; and chose it as the provisional capital of the Territoire Militaire du Niger in 1905, which then spread from Timbuktu and Gao (in today's Mali) all the way to N'Guigmi near Lake Chad.

The colonial administration encouraged migration to Niamey through policies such as abolishing taxes and forced labor in the city, distributing free parcels of land, and establishing major markets while imposing elevated taxes on markets in neighboring towns. By 1911 the population of Niamey reached 3,000. Zinder—an ancient city and the seat of an important sultanate situated 900 kilometers to the east—was made the Chef-Lieu du Territoire Militaire du Niger in 1911. It retained this status until 1922, when Niger became an official French colony. Zinder then served as the capital of Niger until 1926.

In 1926, France moved the capital of Niger from Zinder to Niamey for several reasons. First, France had decided to deal primarily with the Zarma in the administration of the territory of Niger. Second, Zinder lacks an abundant water supply, whereas Niamey rests on the banks of the Niger River. Third, the French and British were still in conflict over territorial and economic issues, and Zinder was too close to Nigeria, by then heavily under British influence, and was too far from Cotonou, the nearest seaport under French control.

Before the French moved Niger's capital to Niamey, the city's residents relied on wells and the Niger River for drinking water. During the colonial period from 1922 to 1960, Niamey saw an elevenfold population rise to 33,816 people (Gado 1997: 50). The colonial administration began planning an urban water regime in the 1920s and 1930s, and it became fully operational in 1940 (Bontianti et al. 2014). Still, residential access was highly spatialized and available only in European neighborhoods located on the plateau. Later, when the piped network was expanded into the *nouvelle ville indigene* (new African city), only public standpipes were available. Private residential access was restricted to European neighborhoods (Bontianti et al. 2014). It is not surprising, then, that water delivery services, such as the one we describe in Chapter 4, emerged to fill the gap in access. While more expensive than going to the public standpipe directly, water delivery saved substantial amounts of time for those would could afford it, and thus mobile water vendors (called *ga'ruwa* in Hausa) became part of the urban streetscape in Niamey.

Niger's peaceful transition from colony to independent state in 1960—and its ongoing reputation for tranquility—contributed to postcolonial Niamey's spectacular growth. The city's population grew to 399,846 in 1988 (Gado 1997: 50). It underwent rapid geographic expansion and became much more densely populated, especially as the percentage of people living in urban areas of Niger increased nearly threefold from 6.22 percent to 15.35 percent from 1960 to 1988 (Motcho 1992: 2). Independence also marks the period during which the Nigérien state took over Niamey's water system.

Several factors influenced the exponential postcolonial expansion of Niamey. Niamey's new status as the capital of independent Niger opened up civil service jobs, and by 1980, 14 percent of Niameyans, or roughly 42,000 people, were government workers (Sidikou 1980: 92–93). Merchants were encouraged to settle. As civil servants and traders established themselves, their families and wide arrays of dependents followed them to Niamey. In addition, Niamey was essentially the only place in Niger where French expatriates settled. Diplomatic missions were established, along with massive international humanitarian and development aid sectors.

The period between the mid-1970s and the late 1980s was a time of relative hope and prosperity. Uranium was discovered in the Sahara of Niger in 1957, and after commercial mining began in 1971 Niger quickly became one of the world's leading producers (Keenan 2013: 93).

Areva, a France-based multinational mining corporation, secured the lion's share of the profits, but enough remained for a modest uranium boom in Niger as a whole and in Niamey in particular. Much of the current skyline of Niamey was constructed with uranium revenues during this period. However, this period ended with uncertainty. Niger was coerced into accepting its first structural adjustment program (SAP) in 1985, jointly organized by the International Monetary Fund and the World Bank (Alidou 2005: 13).

Tensions emerged over the inequitable distribution of the uranium profits and environmental contamination due to careless mining practices, and then the price of uranium on the global market plummeted. This ignited a Tuareg rebellion in northern Niger in 1990 that lasted until 1995 and another from 2007 to 2009 (Keenan 2013). As a condition of its SAP, the government began to sell public enterprises, many of which were purchased by foreign investors, thus draining Niger of potential revenues from these industries, although the water utility company remained under state control until 2001 (Tidjani Alou 2005).

After thirty-one years of repressive single-party rule—first by a civilian regime, then by the military—Niger transitioned to a multiparty electoral democracy in 1991–1992. Democratization occurred partly as a result of local initiatives regarding domestic problems; however, it did not happen in a vacuum. It came in the wake of the collapse of the USSR and the end of the so-called Cold War. Wealthy Western democracies pushed democracy and neoliberal economic policies as preconditions for aid, and a wave of political transitions occurred around the world. Through the introduction of private radio, television stations, and newspapers, democratization allowed a disparate range of voices to publicly articulate their interests, including Islamic groups, women, youth, "rebels," and others with grievances against the state or alternative visions of Niger. The arrivals of the internet (in the 1990s) and cell phones (in 2001) in the midst of Niger's democratic era have further contributed to the exponential acceleration of global and local information and communication exchange.

While many Nigériens were thrilled with the new freedoms of democracy, the majority of the population has experienced economic decline in the past twenty-five years. The 1990s also saw the emergence of a new (to Niamey) water commodity: the hand-tied sachet filled with water. Containing water used only for drinking and washing market vegetables, the sachets were kept refrigerated and sold cold by mobile vendors on the street or in small roadside shops. As Chapter 5 explains, a machine-sealing process has now replaced the hand-tied sachets.

The five SAPs imposed on Niger from the mid-1980s to the present have constituted the primary instruments of the assault of neoliberal economic globalization against Niger (Dearden 2012). While they have helped state balances of payments and perhaps contributed to macrolevel growth, earning Niger praise from the World Bank and the International Monetary Fund (IMF), SAPs have had devastating consequences for the vast majority of Nigériens, and the poorest have been the most adversely affected. SAPs have emphasized cuts in state spending for health, education, staple food subsidies, and other vital social services. SAPs also contributed to inflation, particularly in the cost of staple foods, adding to the decline in standards of living. In addition, "the devaluation of the currency of francophone Africa, the CFA [by 50 percent in 1994], plunged the country into a devastating economic depression with severe sociopolitical and other consequences" (Alidou 2005: 13) from which Niger in general, and Niamey in particular, have never recovered.

SAPs also pushed the privatization of state industries, leading to the selling of key nationalized economic sectors and mass layoffs of Nigérien workers. Although privatization of the water utility occurred later than other sectors, reforms in 2000–2001 led to the creation of SEEN and SPEN (described in Chapter 1) and the complex public-private partnership between the Nigérien state and private companies. During these reforms, Vivendi (now named Veolia Water) purchased a 51 percent share in Niger's water utility. Via SEEN, Veolia is now involved in several elements of Niger's water economy, including but not limited to extending water mains, installing private and public connections, and establishing pricing regulations (Tidjani Alou 2005). Veolia's reach is global, as it is the largest owner of municipal water supplies in the world and was responsible for "upgrading" the pipes in Flint that later contaminated the city's water supply.

SAPs were further challenged by Niger's population growth and growing urbanization, especially in Niamey. Niamey's population tripled from 400,000 in 1988 to 1,222,06 in 2010 (Gado 1997: 50; Hamidou and Ali 2005: 96; République du Niger 2010: 1). New neighborhoods regularly appear, and Niamey's geographic footprint expanded fivefold from 1988 to 2010, from 4,844 hectares to 23,930 hectares, and became even more densely populated (Motcho 1992: 3; République du Niger 2010: 1). Furthermore, the percentage of people living in urban areas of Niger rose from 15.35 percent in 1988 (Motcho 1992: 2) to 20.4 percent in 2010 (République du Niger 2010: 1).

Then, Niamey grew by more than 50 percent in just eight years, reaching almost two million residents in 2018. In a report for the

Ministère de la Population et de l'Action Sociale, Hamidou and Ali (2005: 96) project that Niamey's population will rise to 2.1 million in 2020 and then to 3.7 million in 2030, before reaching 11.4 million in 2050. Some studies indicate that today Niamey is the world's fastest growing city (Demographia 2015; *The Guardian* 2015). Migration from throughout Niger has remained a major factor in Niamey's growth during the 1990s and into the twenty-first century, just as it had in the past. Internal growth, spurred by the world's highest fertility rates of 7.1 (République du Niger 2010: 1), also played an important role.

During most of the years of the past two decades, Niger was rated at or near the very bottom of the United Nations Human Development index. The World Bank and the IMF deserve blame for many, though not all, of Niger's woes. The withdrawal of state support for health, education, and employment in the world's least developed nation as required by SAPs constitutes a neocolonial attack and global crime against Nigériens. However, these powerful financial institutions did not cause Niger's plight alone. A corrupt ruling elite and their merchant allies, the continuing power of the military, the meddling of France, predatory multinational mining corporations, and a difficult environment that has long been prone to drought and appears to be getting worse as a result of recent global heating, have all played a part. Meanwhile, a small but influential and growing minority of Nigériens was positioned to benefit from economic liberalization, leading to sharply expanding inequality between social classes and resentments about it—as is always the case with SAPs in particular and global capitalism in general.

Contemporary Niamey

Today, Niamey has about six times the population of Niger's second and third biggest cities, Zinder and Maradi, which have about 375,000 and 325,000 residents, respectively (INS 2017). Niamey rests atop two plateaus reaching over 200 meters in altitude and is bisected by the Niger River. Only the Kennedy Bridge—named after US president John F. Kennedy—spanned the river from 1970 until 2010, when a China-based firm completed construction of *Le Deuxième Pont* (The Second Bridge). Most activities and at least three-quarters of the population are concentrated at the northeast bank, the *rive gauche* (left bank), of a wide bend in the Niger River. The Communauté Urbaine de Niamey (Urban Community of Niamey) is headed by a mayor and municipal council and includes five urban communes (each with its own mayor) divided into forty-four districts and ninety-nine quartiers—each with its own chef and elected board.

Located firmly within the Sahel region of West Africa, Niamey experiences three seasons, two of which are dry. The cold season, when nighttime temperatures fall to 60°F and daytime temperatures typically reach 90°, lasts from late October through mid-March. The hot season, when daytime temperatures reach above 110°F, begins in late March and continues into June. The relatively short rainy season helps alleviate the intense heat beginning usually in mid-June and lasting until late September. Rain falls every few days, sometimes in intense thunderstorms. Prior to the arrival of seasonal rain, huge dust storms are common, as the winds created by the movement of the intertropical convergence zone and regional pressure cells pick up the dry earth that has not seen rain in more than six months.

Despite its location in the far western reaches of the country, Niamey is the economic and political center of Niger. It is home to several large markets, including the Grande Marché, which is one of the largest formal markets in West Africa. It serves as the main export platform for most of Niger's contributions to global trade, including uranium ore, livestock, and onions. Niamey lacks a strong industrial base, and most consumer goods are imported, as are petroleum and rice. Furthermore, Niger produces little of its own electricity, so residents of Niamey and other regions rely heavily on Niger's connection to Nigeria's electrical grid. In fact, only 62 percent of households in urban Niger are connected to the electrical grid (USAID 2018).

Niamey also serves as the seat of national government. As a democracy, Niger enjoys a multiparty system whereby the elected president serves as head of state and the elected prime minister serves as head of government. A national assembly serves as the legislative branch and the supreme court occupies the judicial branch. Niger has been plagued with several coups d'état in the 1990s, 2002, and 2010. After the most recent one, representative democracy and free elections were restored in approximately a year. The constitution established in 1992 was amended twice, and the current version voted on in 1999.

Still, Niger suffers from government corruption at most levels and lacks consistent long-term national or regional planning, and the striking wealth gap between rich government officials and the severely impoverished population is nowhere more visible than in Niamey. International government organizations, like the United Nations, the World Bank, and the International Monetary Fund control Niger's (and Niamey's) political and economic agenda to their own benefit. The United States has also recently become a major influence on Niger's political agenda, as the Sahara has become a target for the so-called War on Terror. One US drone base is operational in Niamey, and a second one is nearing completion adjacent to Agadez; rumor has it that a secret third base is located near Arlit. More than 800 US Department of Defense (DoD) personnel were present in Niger as of February 2018, the second largest concentration of DoD personnel on the continent, and that number is expected to rise (Penny 2018).

Niamey is simultaneously the most and least representative place of Niger. Because it includes all ethnic groups of Niger, roughly in proportion to their overall national representation, Niamey is, in a sense, highly representative of Nigérien cultures. This same reality also makes Niamey unique and has led to the development of distinct, hybridized, modern urban culture. Niamey also stands apart from the rest of the country due to its size and political influence. Furthermore, the city is far more acutely and directly affected by global trends, policies, and technologies than other places in Niger. While Niamey might not be a "global city" by some definitions, it clearly serves as Niger's most important "gateway to the global world" (Hansen 2008: 4). As "one of the spaces where major macrosocial trends materialize" (Sassen 2007: 100), Niamey offers a strategic site for studying the local faces of globalization.

The sun and searing dry heat are relentless and daunting almost year-round in the Sahel. In Niamey there are, nevertheless, just enough gardens and standing pools of water to create year-round habitats for malarial mosquitoes. When the heat is not oppressive, windblown sand and dust fill the sky and lead to meningitis epidemics and other respiratory ailments. Rain brings temporary relief when it falls sporadically from July to October, but heavy rainfall can cause flooding and the destruction of homes constructed of *banco* (mud bricks).

Despite severe poverty and these climatic challenges, Niamey is a remarkably vibrant, lively, sociable place featuring a distinctive cosmopolitan, public, outdoor culture. The most common jobs—in trading, services, and manual labor—take place outdoors. Most prayers are performed in open-air mosques and in sidewalk gatherings that sometimes expand to block off roads. Most men participate several hours daily in street corner conversation groups. The central districts feature high-rise buildings occupied by banks and government offices; luxury retail shops; a variety of international restaurants; hotels, bars, nightclubs, and casinos; markets; ambulatory hawkers and traders occupying any available space; biomedical, Chinese, and traditional pharmacies; international standard sports stadiums, the national museum, and cultural centers; enormous billboards; and streets clogged with motorcycles, cars, hand-drawn and animal-driven carts, and people—men and women, old and young, from the fabulously wealthy to desperately poor beggars, of all of Niger's ethnic groups and from around the world.

Wide avenues and roundabouts funnel traffic into the mazes of narrow dirt roads of residential neighborhoods. Here the streets are less crowded, the pace is slower: in many ways it feels like rural Niger. Residents typically know their neighbors and greet dozens of them daily. Most live in modest, densely crowded, rectangular compounds constructed of handmade mud bricks or industrially manufactured cement, but a few ostentatious grand villas, round thatch dwellings, and cardboard and plastic shacks are found interspersed in almost every district. Regardless of scale, high walls that face streets and provide residents with a measure of privacy surround most homes. Slightly more than half (54.4 percent) of residences have electricity, while just over one-third (38.3 percent) have running water (Hamidou and Ali 2005: 7). Almost every block has a small mosque or designated prayer space. Men gather in shady spots along the roads to sell roasted meat, serve coffee and tea, and chat. Women—carrying loads atop their heads and babies strapped to their backs—are constantly coming and going to markets, to work, or to visit family and friends. Children are ubiquitous—laughing, kicking soccer balls, and working.

Niamey is probably the safest capital city in West Africa and violent crime is rare. However, underfunded urban infrastructure has not kept pace with growth, and declining economic conditions increasingly thwart the ambitions of many residents. While Niamey has expanded, many Nigériens are migrating out of the country, typically using Niamey as a steppingstone, that is, as a place to gain urban experience and to secure travel documents. The majority of Niameyans of prime migratory age (fifteen to forty-five years) want to leave the country, for a few years at least.

In his book *Surviving with Dignity: Hausa Communities in Niamey, Niger*, Scott Youngstedt (2013) describes Niamey as the lungs of Niger, inhaling migrants, electricity from Nigeria, humanitarian and military aid, and consumer goods from China and elsewhere. Niamey exhales natural resources and agricultural products, substantial amounts of money to wealthy countries to pay off economic development loans, and some of its population who emigrate in search of better opportunities elsewhere. As the political, economic, and population center of the country and home to the only year-round source of surface fresh water, Niamey is essential for the continuation of Niger and is intricately connected through economics, communication technology, transportation, and kinship ties to even the most remote areas of the country. Niamey is also Niger's main connection with the rest of the world. It is within this context that we explore water economies in the city.

3

ACCESSING WATER IN NIAMEY

Hassane (a pseudonym), a 54-year-old Bakourfaye Hausa man, has been living primarily in Niamey since he first came to the city as a seasonal labor migrant when he was a teenager. He spent his youth in Matankari—a village of roughly 1,000 residents about 300 kilometers east of Niamey—absorbing local lore about how to recognize when a rainy season has "set" and hence when planting can begin and helping his father raise millet, beans, and peanuts at the base of towering mesas in the windswept Sahel. Throughout his early years, he drank only well water drawn by hand several times daily by his mother from a communal village well and deposited in *tuluna* (local clay pots made by women; Figure 3.1) in his father's compound.

This difficult labor, considered women's work by the Hausa (and other agricultural peoples of Niger), involved manually pulling leather satchels of water from deep wells, pouring the water into 25-liter clay pots, and carrying that water home atop their heads—about 400 meters for Hassane's mother. Alternatively, women carried water in two 25-liter repurposed petrol cans attached to either end of wooden pole balanced over their shoulders. *Tuluna*—sometimes called "African refrigerators" by Hassane and many other Nigériens—are water permeable and thus they transpire, keeping the water cooler than the air. Like everyone in Matankari, Hassane drank water from small clay cups, recycled tomato paste cans, or hollowed out gourds that he carried to his family's farm. Matankari has long been known for its clean and delicious well water, though it must be subject to periodic contamination due to dirty leather bags and uncovered wells close to livestock.

Hassane first tasted treated, piped water when he arrived in Niamey in the early 1980s and recalled, "I thought it was amazing and that it proved that Niamey was truly a modern city." Like most residents of Niamey, Hassane has relied primarily on the daily delivery of water by *ga'ruwa*, professional male vendors who gather water at public standpipes. He believes that this is great because it relieves Niamey's women

Figure 3.1 Locally made clay pots, called *tuluna*, for sale in a Niamey market. Photo by the authors.

of the difficult labor faced by their rural counterparts of drawing and carrying water. During his initial years in Niamey, Hassane worked an assortment of menial jobs—ambulatory trading, domestic work, and manual labor. By living frugally and through the generosity of friends from the Matankari community in Niamey, Hassane opened, and for the next twenty years operated, a small roadside business from atop rickety tables shaded by a thatch hangar. He sold a range of products typical of small informal traders throughout Niamey, including powdered milk, tins of tomato paste, sugar, tea, candies, chewing gum, kola nuts, cigarettes, and fruit.

Hassane, however, obtained one advantage over many of his rivals when friends from Matankari who run a refrigerator and air conditioner repair shop gave him an old refrigerator, allowing him to sell cold bottles of Coca-Cola, Sprite, and Fanta as well as milk and yogurt drinks in sachets. Thus, in 1990, at age twenty-seven, Hassane had his first cold drinks—a Coke and milk. They were not especially cold, given that his old refrigerator was sitting outside and subject to regular city power outages, but they were decidedly colder than the *tuluna*-cooled water that he experienced up to that point in his life. Hassane rarely bothered with these cold drinks, primarily because they were too expensive for him but also because he had retained the traditional Hausa perception—which he learned while growing up in Matankari—that being cold leads to serious, even fatal, health problems.

Unlike other traders, Hassane never sold cold water in hand-tied plastic bags because he calculated that he made much more profit selling soft drinks and milk. Machine-sealed, labeled one-half-liter bags (or sachets) of cold water—called *piya wata* (pure water)—were just becoming widely available in Niamey when Hassane closed most of his business in 2010. He began occasionally drinking *piya wata* about three years ago, and by 2017 he was drinking about three bags daily when he was away from home during the daytime. He buys this from itinerant vendors who use coolers on pushcarts or small bowls carried atop their heads. Today's *piya wata* is usually far colder—even icy cold—than the drinks Hassane served from his antique refrigerator.

As *piya wata* became ubiquitous in Niamey, Hassane gradually adopted the new capitalist-created idea that this water is more modern and hygienic than piped water and that its purity is due to its coldness and machine-sealed plastic bags. As he put, "I trust that *piya wata* is the cleanest water in Niamey, along with bottled water—but I have never tasted bottled water because it is too expensive for me." Furthermore, unlike most Nigériens his age, Hassane has all of his teeth and good dental health and thus can drink cold water without pain, probably because he never developed a taste for soft drinks and the very popular heavily sugared Nigérien tea and coffee.

Like almost all people in the poor neighborhoods where he works and hangs out with friends, Hassane discards empty plastic bags on the street immediately after consuming their contents. As he explains, "There is nothing else to do with them." Given that municipal waste disposal services rarely cover poor neighborhoods, city streets and sewers are clogged with millions of sachets and other plastic bags, and some of it inevitably ends up in the Niger River. Plastic bags now also cover many trees and bushes in Niamey as well as other Nigérien cities and towns. Sometimes informal trash collectors with pushcarts sweep up the streets and then either burn the trash—which consists largely of plastic—in heaps, emitting highly toxic smoke, or deposit it in vacant spaces and roadsides.

Despite his newfound appreciation of *piya wata*, Hassane is not picky about drinking water. In Nigérien cultures, water is always the first thing that is offered to guests, and Hassane always accepts water when he visits friends and never asks about its origins. At home, Hassane, his two wives and nineteen children, and about thirty neighbors who live in the same compound drink treated, piped water from a single tap that was installed a few years ago. Some of this is stored in *tuluna*, but as Hassane's *tuluna* have broken over time, he has replaced them with plastic bins made in China. When he hangs out with his friends during the day, he does not hesitate to drink water from a plastic motor oil jug that retains black smudges on its interior or from a nearby tap that offers free water drawn from a well built by an Islamic humanitarian organization with which he is affiliated.

Hassane's trips home to Matankari have become more frequent and longer in recent years as he has aged and scaled back his business in Niamey. He explains that in Matankari, "I love the village community, I am content to drink cool well water, and I appreciate that few plastic bags fill our streets and fields." In recent years, women's labor has been eased by the addition of pulleys to village wells and the construction of a few boreholes that draw water into standpipes (Figures 3.2 and 3.3)—though people must pay to use the taps at the boreholes. Hassane, however, worries about the future of Matankari: "The onsets of rainy seasons have become more unpredictable, rainy seasons are shorter than they used to be, and violent storms come more frequently than

Figure 3.2 Women at a well with installed pullies in Matankari. Photo by the authors.

Figure 3.3 Women and children collecting water from a standpipe serviced by a borehole in Matankari. Photo by the authors.

they did in the past." When we accompanied him on visits to Matankari in 2010 and 2017, he and his friends took us to visit a nearby lake that people rely on for fish and the irrigation of year-round gardens—a rare luxury in Niger. However, Hassane pointed out that this is the last remaining lake in the region and recalled that there were several lakes within walking distance of Matankari when he was a child: "I do not know how people will manage here if this one dries up."

This profile of one man's life experience with drinking water in Niger involves many of the key themes of this book, including the diversity of drinking water sources, the commoditization of water and water delivery, the gendered nature of water delivery in urban and rural contexts, inequality of access, changing material cultures of water, folklore, public health, the negative impact of Niamey's *piya wata* culture and economy on the environment, and looming climate change. In this chapter, we survey the diversity of drinking water sources in Niamey, highlighting sources that are not discussed in detail in other chapters.

The water situation—that is, relatively affordable and reliable access to treated water—is better in Niamey than anywhere else in Niger. Nevertheless, and despite Niamey's modern water treatment plant and network of pipes, access to water remains precarious for many residents due to poverty and stoppages in water flows, especially during the hot season. Many international NGOs work on water issues in villages and towns across Niger, but none of them work on water in Niamey (except in offices to plan their rural activities). To fill gaps in access left by the corporate-municipal and international NGO systems, Niameyans rely on many creative strategies, including using piped water directly, buying water from street vendors, sharing water, and obtaining water from their own NGOs and community associations, wells, and boreholes.

Professional Home Delivery of Water

The most common source—at least 50 percent—of drinking water in Niamey is collected at public standpipes and delivered to homes by vendors most commonly known by the Hausa term *ga'ruwa*. We examine the *ga'ruwa*'s lives and work in detail in Chapter 4. About two-thirds of Niameyans lack piped water taps in their compounds. However, this does not mean that all tapless residents get all of their water from the *ga'ruwa* because most also get some of their drinking water from a variety of other sources, which we identify below.

Piped Water in Homes

About one-third of Niameyans enjoy piped water taps in their compounds, but this aggregate figure obscures the diversity of the ways they procure water. In many compounds—such as Hassane's, which includes six rental units—dozens of people share a single outdoor tap. These can periodically run dry, especially during the hot season, compelling residents to seek water from other sources. Furthermore, compounds that fall behind in paying their bill have their taps locked by SEEN workers.

Only a small minority of households can afford one or more indoor taps. However, this treated water often contains sediment, depending upon the age and condition of pipes. Hence, those with sufficient financial resources install filters on their taps. Wealthy households also use water storage tanks—the largest of which are placed on rooftops or atop their own towers—as insurance against stoppages in water flows.

Sharing Water

Sharing is the key to survival in Niamey and across Niger. In Niamey, most adults are either unemployed or underemployed, and public sector work has been gutted due to neoliberal policies, such as structural adjustment programs. Most Nigériens rely on generalized reciprocity, that is, dense networks of continuous sharing and gifting. This includes remittances sent home by international migrants and rural-urban migrants within Niger and the rotating credit schemes operated by *fada* (men's associations) and *foyandi* (women's associations). However, it is simple everyday sharing that is the staff of life in Niamey. For example, when an adult is sitting down for a meal—even if it is the only meal he or she can afford that day—and is approached by a family member or friend, he or she will typically say, *Bismillahi* (In the name of God, please be my guest). And usually Nigériens really mean it, or at least feign offense if their offer is refused.

This ethic of sharing, of course, includes water. Water is always the first thing offered to guests in Niger, as already discussed. Men visiting their favorite *hira* (conversation groups) typically purchase bags of *piya wata* to distribute to their friends. The gifting of water also occurs on a larger scale, particularly when wealthy neighbors give water to poor neighbors. This practice probably occurs most frequently in upper middle-class neighborhoods such as Dar es Salam, where we lived for nine months in 2016–2017. Very few *ga'ruwa* work in these neighborhoods since most homes enjoy indoor tap water. Nevertheless, about 10 percent of residents in upper middle-class neighborhoods are squatters in vacant lots facing a precarious position because they do not have running water in their homes and may not have any *ga'ruwa* servicing their area. In response, some wealthy men run pipes through holes in their compound walls to offer free tap water on the street for their neighbors. In other cases, ours included, families living comfortably allow squatters to enter their compounds to fill up *bidons* from outdoor garden taps. In these ways, some Niameyans obtain most or all of their water for free.

Sachet Water

Local entrepreneurs introduced manually filled and hand-tied one-half-liter plastic bags of water in the 1990s, as mentioned briefly in the introduction. Machine-sealed bags of sachet water hit the market in the early 2000s and gradually replaced the practice of manually sealing bags, though the latter is still used for selling ice. We examine the lives of those who produce, sell, and consume this fastest growing source of drinking water in Niamey in detail in Chapter 5, and in Chapter 6 we discuss branding.

Household Collection at Public Standpipes: The Case of Sodja Pompo

Some households forgo the *ga'ruwa* delivery system and collect water themselves, particularly if they are destitute but are fortunate enough to live very near public standpipes (Figure 3.4). They must pay standpipe managers, but this is less expensive than paying *ga'ruwa*, who must also pay standpipe managers for the water that they collect. This involves difficult work since the standard 25-liter water containers weigh twenty-five kilograms each (or fifty-five pounds). Many very poor families do not own wheelbarrows or pushcarts, and thus must carry the water by hand. This task is typically carried out by children

Figure 3.4 Children collecting water in *bidons* directly from a public standpipe in Niamey. Photo by the authors.

and youth, who must make about ten trips daily to standpipes to collect enough water to accommodate the drinking, cooking, bathing, and ablution needs of typically sized households.

To illustrate the conditions involved in using public standpipes as a water source, we describe the situation at a public standpipe in the peripheral neighborhood of Sonuci. In 2016 and 2017, we made weekly visits to the public standpipe, at different times of the day or different days of the week, sometimes spending several hours observing activity surrounding the public standpipe and talking to Sodja, the standpipe manager.

Sodja Pompo—a Hausa-Kanuri man—was born in 1958 in a village near N'guigmi in the southeastern corner of Niger. Everyone in the Sonuci neighborhood where he lives and works knows him as "Sodja"—a term of respect—because he served as a soldier in the Nigérien military for more than thirty years before he retired in 2008. (*Sodja* is a Nigérien Hausification of the English word "soldier.") His second nickname, "Pompo," identifies him as a standpipe manager— *pompo* is a Nigérien Hausification of the English word "pump." Unlike many of Niamey's peripheral neighborhoods, Sonuci is a carefully planned formal neighborhood for the wealthy with wide streets and specifically demarcated residential lots laid out in a grid pattern.

Sodja spent his military career in western Niger, including Niamey, due to the longtime Nigérien state policy of sending civil servants far from home as a strategy for building national unity. Planning for his retirement, Sodja purchased a small lot in Sonuci and built a modest two-room cement home with electricity and running water—roughly middle class by Nigérien standards—in 2003. After retiring, he and his wife moved into the home and he became the neighborhood standpipe manager in 2009—a time when, as Sodja put it, "There was nothing much here but empty lots."

Most standpipe managers are Tuareg and Fulani, but there are two good reasons a Hausa-Kanuri man took the job. As a result of a career spent in western Niger and marriage to a Zarma woman, Sodja now speaks Zarma better than either Hausa or Kanuri and sometimes identifies as Zarma—a prime example of the common practice of *brassage*, or "ethnolinguistic mixing," in Niger. First, Sodja was one of the earliest residents of the neighborhood and by chance the first standpipe in the neighborhood was located just outside his home. Second, *ga'ruwa* do not work in Sonuci because most residences have indoor tap water, and hence there are few Tuareg and Fulani in the neighborhood.

Sodja is the most well-known standpipe manager in Sonuci— and perhaps in all of Niamey—even though his standpipe has few

Figure 3.5 "Sodja Pompo" sign with standpipe in the background in Sonuci neighborhood, Niamey. Photo by the authors.

customers. He is usually very humble, but he proudly told us, "All of the taxi drivers in Niamey know me and where I work"—which we confirmed by sometimes taking taxis from distant parts of the city to visit him. This seems to be due to his engaging conversational skills as well as the fact that he has a small sign next to his standpipe with just two words, "Sodja Pompo" (Figure 3.5). We did not observe signage at any other standpipe in Niamey.

This relatively new residential neighborhood includes many multistory expansive villas and very little commercial activity other than a few informal boutiques, mechanics, and meat-grilling stands. Sonuci's unpaved boulevards are wide enough for stockmen to walk herds of 100 cattle from the bush—only about 300 meters from Sodja's standpipe—to a small pond in the neighborhood. Roughly 20 percent of the lots include partially completed homes or ones where formal construction has not yet begun. Squatters occupy many of these lots, living in thatched circular homes. Since there are not enough of them to make it worthwhile for *ga'ruwa* to work in Sonuci, squatters go to Sodja's standpipe or one of the other two standpipes in the neighborhood. His other customers are construction workers.

In contrast to the nonstop activity during daylight hours at standpipes in densely crowded poor neighborhoods near the city center, the customer flow at this standpipe is slow. Customers—who pay 15 FCFA (West African franc) for twenty-five liters of water—arrive about every fifteen minutes during the hot season and less often during the cool season, typically with two 25-liter repurposed cooking oil containers on wheelbarrows. Others bring six to ten containers at a time on donkey-pulled carts (Figure 3.6). Sodja rarely leaves his standpipe, and when he does it is usually just to enter his compound, where he can be reached with a knock on the door, or to pray at a nearby mosque from which he can still see customers. Customers know this and wait for his return or drop off their containers and return later.

Sodja explained, "I must work, it is better than sitting and doing nothing." He spends most of his time chatting on wooden benches with friends on his wide street corner. Sodja usually spends a few minutes greeting and chatting with customers. He insists on filling the containers himself, and despite his slight frame lifts them onto carts for children and old men and women, belying his comment, "I am too old to do anything else, and I am not so strong anymore." (We once saw

Figure 3.6 Sodja helping a customer fill *bidons* at his standpipe. Photo by the authors.

Sodja—with the help of one friend—strap his full-sized refrigerator to the back of his small motorcycle. He then rode several kilometers along sandy streets to a repair shop—a task that would challenge even a strong young man.) Sodja enjoys working and socializing at the center of this hydrosocial territory, even though he earns very little money as a standpipe manager and relies primarily on his military pension. He likes the responsibility, including traveling about five kilometers once a month to pay the SEEN bill, and a small sense of power derived from the fact that he owns the key to the standpipe and could deny water access to his neighbors (though he has never done so).

The water never stops flowing from Sodja's standpipe, even during hot seasons, because it sits directly over the Lazaret water main and the pipe system was designed to service a wealthy neighborhood with private indoor taps. Sonuci's reliable water service is a classic example of investment patterns in urban water services in developing countries. Although it is one of the newer neighborhoods in Niamey, Sonuci contains water infrastructure because the chances are good that most households will establish direct connections to the piped network (which many have) and have the means to regularly pay their water bills. Thus, the opportunities for recuperation of investment are good. Furthermore, many poor neighborhoods between Sonuci and the city center have tenuous water access—an illustration of the archipelagos of access (Bakker 2010) that characterize the waterscapes of Niamey.

Wells, Boreholes, and the Niger River

Wells and boreholes may provide clean water, but there are no guarantees because these sources of water are not chemically treated. A few Niameyans, particularly in neighborhoods on the outer periphery of the city that until recently were villages, rely on traditional, hand-dug, manually operated wells. Probably a few continue to operate within the city behind compound walls, but none of the people we interviewed mentioned the use of wells within the city. Boreholes have become a popular alternative for the wealthy and for community associations that can pool resources, as discussed further below. Once the borehole is operational, the owner does not need to pay for water and anticipates that over time this will offset the expense of installing the borehole. Particularly destitute and mentally ill Niameyans sometimes drink water from the badly polluted Niger River. Those living close to the river use it for bathing and washing clothes, and swimmers, bathers, clothes washers, and others must inadvertently ingest some river water.

Community Associations: The Case of Kituba was Sunnah

Many local community associations in Niamey offer humanitarian social services, including the provision of free drinking water. The following case study of one association offers an example of these activities.

Elhadji Mamane (a pseudonym) led a mosque and community in Abidjan-Kalley Sud—an old, mostly poor neighborhood not far from the city center—for thirty-five years until his death in 2017. As his reputation as a famous learned and generous *malam* ("Muslim scholar-teacher") grew over the years, he drew other well-known *malamai* (plural of *malam*) into his congregation to study, preach, and teach and garnered financial support from several wealthy businessmen in the neighborhood. Elhadji Mamane's mosque and school are attached to his sprawling yet humble compound, which always offered guest rooms, food, and water to visiting *malamai* and guests from his home village—some of whom stay for months or even years at a time.

For thirty years he organized *wa'azai* (public sermons) on Thursday nights—the traditional time for *wa'azai* in Niamey. Elhadji delivered many *wa'azai* until about a decade ago, when a mysterious throat ailment robbed him of his once powerful voice. These elaborate gatherings draw audiences of several hundred people, last about three hours, and usually involve two to four *malamai*. They involve blocking traffic from entering the large public space in front of Elhadji Mamane's house, setting up and taking down wires and speakers, and rolling out and rolling up mats that are used to accommodate at least 100 of the most honored attendees.

Elhadji Mamane's congregation—like many or most in Niamey—has grown more conservative over the years. Moving away from their "traditional" Nigérien loosely Sufi Muslim roots, about twenty years ago they began identifying with the Salafi-inspired Izala movement, which originated in Northern Nigeria and has ties to organizations in Saudi Arabia. A few years ago, Elhadji Mamane and his colleagues established direct links with and began receiving funding from a Saudi Arabian Islamic association, and they began calling their community "Kituba was Sunnah." (In this sense, they are now an international NGO.) This new financial infusion, together with the continuous cultivation of local donors, has been used primarily to expand the congregation's public outreach initiatives in three ways.

First, Kituba upgraded their sound equipment and purchased a new raised speaker's stage for their *wa'azai*. Furthermore, they purchased a station wagon and two vans to transport *malamai* to and from *wa'azai* across Southern Niger and Northern Nigeria. Second, Kituba expanded its food distribution to the poor. This includes regular distribution of food as well as purchasing, slaughtering, and giving to poor residents dozens of sheep and 200 cattle during last year's celebration of Sallar Layya (also called Babbar Salla in Hausa, Tabaski across West Africa, and Id el-Kabir in Arabic.) Third, and of most relevance to the story of water, in 2014 Kituba purchased the compound just across the street from Elhadj Mamane's and installed a borehole.

Kituba bought the compound to serve as a guesthouse for visiting *malamai* and students. For many years it had been occupied by a family that had largely kept to themselves and had not participated in the mosque community or other social activities in the neighborhood—very unusual behavior in a popular neighborhood such as Abidjan/Kalley Sud. The family also owned a big dog that occasionally escaped briefly from the compound and terrorized neighbors with his menacing snarls—also quite out of the norm in poor neighborhoods. Kituba renovated the interior and added two private, outdoor cement buildings with porcelain squat toilets. Finally, and at great cost, Kituba paid for the digging of a borehole and water tower with pipes connected to a public-private tap on the street offering free water (Figure 3.7). Altogether Kituba spent 36 million FCFA (US$72,000) to renovate the compound, dig the borehole, and purchase the water tower and pipes for the roadside taps. This included some contributions from the association in Saudi Arabia as well as several donations of 3 to 5 million FCFA (US$6,000 to US$10,000) each by Elhadji and other local *malamai* and businessmen.

Today, about 100 children and youth in the neighborhood come daily to collect water, one or two 25-liter *bidons* at a time, which they struggle to get home in wheelbarrows on sandy unpaved streets. The tap is also used on a daily basis and during *wa'azai* to fill small receptacles—especially plastic teakettles made in China—with water for ablutions. We were surprised that the area under the tap was not treated with more care. There is no cement basin or drainage system as are found at corporate-municipal taps from which *ga'ruwa* collect water, and since the street here is unpaved it can get muddy very quickly. Some people carelessly gather water or drink directly from the tap, allowing a lot of it to spill, attracting trash (especially plastic), Niamey's roving goats and sheep (and their feces and urine), and flies. Fortunately, youth from the congregation clean the mess about once every ten days. About a year after the compound was renovated and the borehole and water

Figure 3.7 Kituba was Sunnah borehole and water tower across from Elhadji Mamane's compound. Photo by the authors.

tower were constructed, a member of the congregation began operating an Islamic medicine shop just outside the compound's walls.

The water tower is Kituba's largest and perhaps most visible projection of material culture, branding, and advertising (see Figure 3.7). (Kituba's vehicles also identify themselves with painted logos in

Arabic, French, and Hausa.) The tower rests at the front edge of the compound, and since it is about fifteen-feet wide and its scaffolding makes it about thirty-feet tall at its apex, it is clearly visible to passersby on the fairly busy streets below it. It features primarily Arabic lettering and also the words ORGANISATION POUR UN DEVELOPPEMENT HUMAIN (ODH; Organization for Human Development). In contrast, Elhadji Mamane's mosque and school have no signage. Furthermore, four loud speakers rest atop the water tower to amplify the daily call to prayer from Elhadji's mosque and Thursday night *wa'azai*.

Kituba view this offering of water to the neighborhood as an important part of their Islamic humanitarian duty. While Kituba is a proselytizing organization, offering free water is not directly part of that mission. They do not monitor or check the Muslim credentials of people who come to help themselves to water from their tap. Just weeks before he died, Elhadji Mamane, reflecting on his achievements especially in recent years, told us, "In a small way I have helped to bring the healing powers of water and Islamic medicine to the community." He added, after pausing for a bellowing laugh, "And the awful dog is gone!"

Kituba's tap, however well intentioned, has some real and potential drawbacks. It has led to the loss of business for local *ga'ruwa* who delivered treated water in the neighborhood. In contrast to corporate-municipal water, borehole water is untreated, and it is not clear that Kituba's water is tested regularly or at all for cleanliness. Furthermore, according to hydrogeologist Daniel Saftner (personal communication), it is never a good idea to build a borehole close to latrines, and latrines are always nearby in densely crowded, old, poor neighborhoods of Niamey.

Bottled Water

Bottled water is by far the most expensive source of drinking water in Niamey. It is consumed primarily by the very wealthiest of Niamey's residents, though some middle-class families purchase it for their young children. The vast majority of Niameyans have never tasted bottled water. Despite its relatively exorbitant cost, enough demand exists in Niamey to support at least six Nigérien bottled water companies: Belvie (established by immigrants from India), Dallol, Diago, Rharous, Tasnim, and Telwa. Brands from neighboring West African countries are also available. In addition, higher end markets sell bottled water imported from France, such as Eau Vitale, Evian, and Perrier. We return to the topic of branding and labels in Chapter 6.

Though we should have known better, in many previous visits to Niamey we relied heavily on bottled water, assuming that it was the safest source of drinking water. In 2016, we were informed that US Embassy personnel had tested all Nigérien brands of bottled water, found contaminants in all of them, and recommended against drinking them. French brands, in contrast, passed their health inspections. However, a great deal of evidence indicates that treated tap water is generally safer than bottled water because tap water is more tightly regulated and inspected by governments.

Niameyans reuse drinking water bottles—sold in 0.5- and 1.5-liter sizes—unlike plastic sachets, which are simply discarded, having no further use. Bottles are refilled with traditional liquid medicine, water, milk, and various popular juices, such as baobab, bissap, and ginger.

Pricing Water

In Niamey, the poor pay more for water than the wealthy, a point we have already made. Understanding the pricing structure of water access in Niamey helps us understand how this situation emerged. At the national level, SPEN is in charge of water management in urban areas as a national resource. To connect to the piped network, an initial connection fee is charged as well as a monthly service fee for renting the required meter installed in the compound.

In Niamey, and other urban areas in Niger, the cost charged to residential customers involves a classic inverted or increasing block tariff (IBT) structure. In other words, residential customers pay prices that depend on the level of household consumption; typically, low consumption results in lower amounts charged. This structure is referred to as "inverted" because the standard economics of water provision show that, typically, the cost to the water service provider per unit of water declines as water consumption at a specific connection increases. For the purposes of pricing, however, this relationship is reversed—when household consumption increases, the household moves into a higher tariff range; thus, though the cost of providing water declines, providers profit more. The upper limit of the lowest tariff block is referred to as the "lifeline block," whereby the rate per unit of water is actually lower than the cost to supply it. This format theoretically ensures that the poor can afford basic water needs (Nickson and Franceys 2003).

As home renters in Niamey in 2016–2017, we had the opportunity to directly participate in this pricing structure. Upon arrival in Niamey for the year, with some work, we secured a modest, middle-class home to rent. This house had two bathrooms, each with a flush toilet, sink, and shower, and one half bathroom with just a toilet and sink. We also had a sink in the kitchen and a private standpipe in the yard, from which our guards used water to water plants, wash the car and our clothes, and share with our neighbors. Each month, a SEEN employee in uniform knocked on the door of our compound, read our meter, recorded his visit by writing the date in marker on the metal door to our compound, and later left a bill that we had about two weeks to pay. Most months, our water usage placed us in the second lowest tariff block. As water researchers, we were intimately aware of water access and provision challenges in Niamey in particular, and the Sahel in general, and we were conservative with our water consumption.

However, one month, we received a water bill that placed us in a much higher tariff block. When we compared this bill to previous ones, we saw several things. First, while meter reading had occurred approximately every thirty days before, the bill in question indicated that almost six weeks had passed since the last meter reading. Since the period covered by the bill was longer, our consumption was higher, even though our daily consumption had not changed significantly. Thus, the total amount of water consumed since the last reading was higher than normal and put us just into the next tariff block. In other words, we paid as much for water that month as someone who consumed almost twice the amount, but the block tariff structure lumped us into the same category.

As Bardasi and Wodon (2008) note in their study of water service costs in Niger, IBTs result in increased costs for individuals sharing a connection with several households. The situation in Hassane's compound that we described at the beginning of this chapter is a classic example. In their study, Bardasi and Wodon found that in urban Niger, four is the average number of households living together and sharing a connection to the piped network. So, even if an individual household's consumption is low, the price the household pays for water may be higher because the total of all the households' consumption is higher.

Public standpipes are assigned the rate associated with the lowest consumption block (the lifeline block). However, in cities like Niamey, where more than half of the residential population lacks a direct connection to the piped network, the assumed advantages of IBTs are often not realized (Nickson and Franceys 2003) because the poor prioritize water provision among other household expenses by paying higher

Table 3.1 Cost of Water in Niamey by Transportation Method.

Transportation method	Average cost per liter in FCFA francs (500 FCFA francs = US$1 in 2015)
Piped water in home	0.18*
Direct from public standpipe	0.60
Delivered to home from public standpipe	1.40
Sachet water (sold in half-liter bags)	50.00
Bottled water	500.00

Source: *From Bardasi and Wodon (2008). All other figures derived from the authors' research.

prices to private water providers (PWPs) like *ga'ruwa*, which we explain further in the next chapter.

In addition, Bardasi and Wodon (2008) found that water consumption (per household and per capita) was five times higher in households with a direct connection to the piped network than in households using alternative water providers, presumably because alternative sources of water are more expensive. Thus, the poor are paying more for water than the wealthy. Table 3.1 lists the average price per liter of water from each of the main water sources in Niamey, although water sharing is not included because of the difficulty in establishing the value of social capital. The fact that the poor pay more for water than the wealthy, but use much less water, is one reason advocates for pro-poor urban water policies (see Nickson and Franceys 2003) promote the reduction in connection costs and fees for poor residents.

Conclusions

Niamey's residents use multiple and fluid strategies to access water. Indeed, Hassane's experience—discussed at the beginning of this chapter—is typical of the poor in Niamey. He obtains drinking water from all of the sources described in this chapter, except for bottled water and the Niger River, sometimes in a single day. Niameyans must adapt spontaneously to hard-to-predict scenarios: their taps may cease to flow due to lack of water or failure to pay bills; their local standpipe manager may fail to pay SEEN bills, leading to the closing of standpipes and the loss of *ga'ruwa* services; wealthy neighbors sometimes stop sharing water; and traditional wells can collapse or run dry. Risk of contamination exists for all of these water sources. This tenuous access is a public health menace. Sources of contaminated water are nearly impossible to trace, particularly if people are regularly drinking from multiple sources, and Niameyans may continue drinking polluted water from the same sources.

Advantaged Niameyans enjoy far more secure access to treated water than their poor neighbors. However, even families that can afford to build swimming pools and fill them with water can still face lack of water in their taps, particularly during hot seasons. Those who can afford it and have the foresight to invest, purchase large water containers and install them on towers or on the ground to guarantee continued access to clean water when the corporate-municipal water distribution system fails.

While health risks and variations in availability are real problems for Niameyans, the provision of water through household delivery and sachets is often the most efficient, time-saving, and realistic option for many households. These two forms of water access involve complex sociocultural relationships between producers, vendors, and consumers, and the materials involved in these economies carry with them different perceptions and meanings. The next two chapters explore in detail the economies of household water delivery (Chapter 4) and sachet water (Chapter 5) in Niamey.

4

WATER DELIVERY VENDORS IN NIAMEY

When we seriously began studying water access in Niamey in 2013, we realized we needed a gatekeeper to help connect us with the community of water vendors in some of Niamey's neighborhoods.[1] Our old friend Hassane, knowing about our research, offered to help us meet some *ga'ruwa*. He suggested visiting a standpipe around lunchtime because it is the only consistent time that *ga'ruwa* take breaks, and so we did. We had walked a little over a kilometer from Kalley Sud into Kalley Nord when we spotted a group of four Fulani *ga'ruwa* resting and preparing lunch. Hassane and the Fulani men began their encounter by teasing and insulting each other in a playful way because Hassane's people—Kourfeyawa Hausa—and Fulani have long enjoyed joking relationships. We then greeted the men, informed them of our study, and asked them if they would be willing to talk to us about their work. They agreed but paused for a few minutes to communally share a bowl of macaroni topped with a savory tomato sauce.

The four Fulani men—Ali's group, all in their twenties and thirties—informed us that they were from the bush near Birnin Konni, about 400 kilometers east of Niamey. They had been in Niamey working as *ga'ruwa* for two to four years each. These four wiry strong men weighing about sixty kilograms each perform their difficult labor relentlessly. They work seven days a week from sunrise to sunset, pushing carts with ten 25-liter repurposed cooking oil containers of water each on Kalley Nord's mostly unpaved roads to deliver water to compounds by pouring it into storage receptacles. One of the men explained, "This is hard work. Every day I have pain in my heart, chest, rib cage, legs, and shoulders. I would like to quit, but at least we are in this together and help each other."

On our third visit to the group a few days later, we found only one of the men, Ali, and he was leaving the standpipe without his pushcart. After exchanging greetings, he informed us that the standpipe had been locked and that he had not yet figured out what to do next. Later that day, we located another standpipe—the one Idrissa's group used—about one-half kilometer away. Ali was there. He explained that his team pays a standpipe manager, who then is supposed to pay the SEEN bill. However, the manager had not paid the bill, and hence SEEN locked the standpipe. After Ali described this problem, Idrissa, the manager at this other standpipe in Kalley Nord, agreed to let the Fulani team use his standpipe, even though it was already very heavily used and would mean that *ga'ruwa* would have to wait longer for their turns to fill up with water. Ali was grateful for this gesture but told us that he would have to travel much farther to provide service to his clientele than he previously did from his home standpipe.

Idrissa's informal *ga'ruwa* association includes eight men with a core group of four Aderawa Hausa—including Idrissa—from a village in the Tahoua region of Niger. Three of the four Aderawa Hausa men return home for three to four months each year to farm, always leaving one man in Niamey to retain the standpipe manager position. Idrissa, about fifty-five years old, is the oldest man in the group and has served as the manager for the past couple of years after working as a *ga'ruwa* for twenty years in Niamey. Unlike other managers, Idrissa works about one-third of the time as a *ga'ruwa*. He explained, "I do this to earn a little extra money and to stay strong." Four Malian immigrants round out the group, two Tuareg and two Fulani.

Unlike the Fulani men of Birnin Konni, none of the men in Idrissa's group complained about the physical difficulty of their jobs. Instead, they identified seasonal variation in water flow from their standpipe as their key challenge. Idrissa described this problem succinctly: "The irony of our work is that there is plenty of water in the cool season when customer demand is low but not enough water in the hot season when customer demand is high."

The eight men had established a strong rapport after several years of working together. While waiting their turns to refill their containers and during breaks, they enjoy constant bantering, teasing, and spraying water on each other. They do not have a formal association, but they emphasized the critical importance of solidarity in their work and helping each other many ways, such as covering for men when they are sick, training apprentices, and absorbing new men into the group. When the four displaced Birnin Konni Fulani men were out delivering water, Idrissa explained that his group listened carefully to their story, evaluated their character through a long conversation, and decided that they were good men in a difficult situation.

One week after the Fulani men had been displaced, two moved to work at another standpipe even farther away from their hydrosocial territory. They were not asked to leave but felt they were burdening Idrissa's group. One month later, their standpipe was still locked. They continued to discuss the unscrupulous actions of their manager, exhibiting no discernable anger, only resignation.

Ga'ruwa in Niamey: An Introduction

Most residents of Niamey rely on daily deliveries of water made by ambulatory vendors because only a minority of Niamey residents enjoys running water piped directly inside their homes or courtyards. These vendors—called *ga'ruwa* in Hausa (literally "there is water") and the most commonly used term in Niamey—purchase water at streetside standpipes, where they fill ten to fourteen twenty-five-liter plastic containers of water and deliver it for a higher price to regular customers, using metal pushcarts with bicycle wheels specifically designed for this purpose by local blacksmiths (Figure 4.1). *Ga'ruwa* is a complicated word. It literally means "there is water," although it does not appear

Figure 4.1 *Ga'ruwa* at a public standpipe in Niamey. Photo by the authors.

in Hausa dictionaries. *Ruwa* usually means water, but its meanings include juice, semen, usury, and business among other meanings that are context dependent.

Most *ga'ruwa* in Niamey are Tuareg and Fulani men. Many are immigrants from Mali, especially the Gao and Timbuktu regions, but also from Gossi, Douanza, Hombori, Bori, and Bandiagara. A few are from the far western regions of Niger near the Mali border. At least 70 percent of *ga'ruwa* in Niamey are Tuareg, specifically "Black" Tuareg. "White" Tuareg rarely do this work in Niamey. Many scholars have discussed the complexities of the construction of "race" among the Tuareg (Diallo 2016; Hall 2011; Lecocq 2010). Diallo (2016: 43) explains that in Mali the "terms 'Bellah' and 'Iklan' [and Bellah-Iklan'] are used to describe former Tuareg slaves known as the black unfree Tuareg in contrast to the former masters known as freeborn and 'white' (or 'red') Tuareg." Many of the Tuareg *ga'ruwa* that we interviewed—similar to Diallo's participants—emphatically refer to themselves as *peuple noire* (black people) to distance themselves from "white" Tuareg, who they regard as greedy, lazy, evil, immoral oppressors. In contrast, only a small minority of *ga'ruwa* positions in Niamey are filled by Nigérien Hausa and Zarma, even though together they constitute more than three-quarters of Niameyans.

This chapter offers answers to two related questions. First, why are *ga'ruwa* jobs in Niamey dominated by immigrant Tuareg and Fulani men from Mali? Second, why do Nigérien men in Niamey, particularly from the traditionally sedentary Hausa, Zarma, and Songhay, avoid this job, even though most Nigérien residents of Niamey are underemployed or unemployed? Water delivery is not a great job and requires physically strenuous work, but it is a steady job that offers an above average income. Answers to these questions require an understanding of ethnic identities, interethnic relations, and gender relations among both traditionally nomadic and traditionally sedentary people of the Sahel.

This chapter examines water and its commoditization as a cultural symbol and mechanism for interpreting social relations in the urban context of Niamey. Focusing on the lives and work of water vendors allows us to develop a holistic understanding of the material, symbolic, and cultural elements of water in Niamey in particular and Niger in general. That is, we explore a dialectical relationship. We argue, on the one hand, that *ga'ruwa* offer a conduit for understanding key cultural symbols, values, and social relationships and, on the other hand, that we must understand symbols, values, and social relationships in order to understand the *ga'ruwa*. The *ga'ruwa* are ideally positioned to explore this dialectical relationship because they do not operate within an established infrastructure.

As this chapter shows, they are infrastructure (Simone 2004). In his key paper, "People as Infrastructure," Abdoumaliq Simone (2004: 409) explains that in Africa "state administrations and civil institutions have lacked the political and economic power to assign the diversity of activities taking place within the city (buying, selling, residing, etc.) to bounded spaces of deployment, codes of articulation, or the purview of designated actors." These conditions influence how people live, negotiate, and collaborate within the urban context (Simone 2004: 410). It is within the context of state negligence and the inability of residents within African cities to improve their livelihoods that the water vendors function.

Tuareg and Fulani of the Sahel and Sahara have endured crises for generations that have threatened and disrupted what they regard as central to their identities and as the ideal, most dignified way of life: nomadic pastoralism. These crises include colonial conquest, postcolonial droughts, state neglect, rebellions, and terrorism. Tuareg and Fulani have responded to these challenges by creatively adapting various complex "kinship, social, and trade networks that they have formed over time across ethnic boundaries, empires, and nations" (Giuffrida 2010: 23). We draw on Alessandra Giuffrida's (2010: 23) "integrated approach to mobility and stasis," which defines "mobility and stasis in systemic terms beyond pastoralism … to consider interrelations between different categories of mobility and networks … [revealing] … structural fluidity and change in contemporary Tuareg [and Fulani] societies." Giuffrida (2010: 24) identifies "intensive and extensive pastoralists, returned migrants and refugees, seasonal rural migrants and temporary urban migrants" as categories of mobility and stasis that constitute a system of adaptive strategies.

Loftsdóttir (2008: 137) similarly characterizes this diversified economic adaptation as a complex strategy of risk avoidance akin to a broad portfolio of investments. This approach de-essentializes nomadic-sedentary and rural-urban divisions through emphasizing local and translocal social ties. Thus, the nomadic and the sedentary are linked through remittances and circular labor migration.

New fluid identities are emerging among Tuareg and Fulani that incorporate nomadic pastoralism as well as experiences in diaspora, including a new relationship with water in their role as *ga'ruwa*. Today, there are very few pure nomads; that is, those who do not rely on the support of urban kin and friends or who have not spent some time away from their herds as urban labor migrants or refugees. Mobility and stasis are "complementary rather than in opposition" (Giuffrida 2010: 30). Despite this interdependence, balancing multiple identities presents many challenges.

Among Tuareg, people of the bush "who have never left home see their returning relatives as corrupt and impure because of their exposure in cities and refugee camps to foreign values, diets, beliefs and behavior" (Giuffrida 2010: 32). In contrast, some returning and visiting migrants, particularly if they have had some success in the city, "view pure nomads as poor, backward and trapped in a miserable existence. Nevertheless, they exalt their ability to endure such an existence because if they did not their heritage and cultural identity as Tuareg would disappear" (Giuffrida 2010: 32). Similarly, most urban Fulani desire to return to the bush, because for them:

> The bush is sweet as sugar, while Niamey is the place of corruption; the bush is a place of freedom for individuals while in Niamey they are constantly being observed; in the bush people eat food that makes them strong and healthy, while in Niamey they eat food that lacks power and is unhealthy; in the bush people are surrounded by family while in Niamey they are without their closest kin. (Loftsdóttir 2008: 139)

However, as Loftsdóttir (2008: 139) explains, "Some of the same people who previously [while living in the city] had idealized the bush complained [upon return] about how hard and dull the life in the bush was most of the time." As a result of circulating stories about these experiences, "some Fulani who have been able to gain the means of returning to [the] bush have chosen not to do so, but have preferred to stay in the city, envisioning going 'back' in an unknown future" (Loftsdóttir 2008: 140).

We continue this chapter, first, with a detailed ethnographic examination of water vendors in Niamey. Then, we examine the emergence of the *ga'ruwa*, the process of becoming a *ga'ruwa*, the challenges faced in this work, and *ga'ruwa* interactions with standpipe managers and water bureaucracies. In the second section, we discuss the critical importance of solidarity and cooperation among the *ga'ruwa*. In the third section, we consider the symbolic value of water and the social relationships that revolve around it, particularly gender and ethnic relations. Finally, we conclude by offering answers to our preliminary questions. Underlying these sections and drawing from the general framework utilized in the recent interdisciplinary volume *The Social Life of Water*, edited by John R. Wagner (2013), we explore ways by which the triad of urbanization, technology, and commoditization affect access to and the quality of water as well as the lives of people involved.

Water Vendors in Niamey: Exploring the Role of the *Ga'ruwa*

One factor driving the expansion of Niamey's piped water network was the growth of the city's population. Today, the state and the private sector compete for profits, distribution, and regulation of urban water policy in Niamey. Niameyans have had to pay for water since formal distribution structures were established. Over time, drought, climate change, and increasing rates of rural to urban migration have encouraged the emergence of informal paths to water access, such as mobile water vendors, alongside these formal structures of water distribution.

Hybridized processes of water access are not unique to Niger. As we described in Chapter 3, many residents of African cities use a multitude of networks to access water, especially when treated, piped water is expensive and inconsistent in its provision. These include, but are not limited to, neighbors, pirate provisioners, and private vendors (Myers 2011). Several studies consider water vendors in other African contexts, namely in Ghana (Osumanu 2008; Osumanu and Abdul-Rahim 2008), Kenya (Whittington et al. 1989), Sudan (Cairncross and Kinnear 1991), and Tanzania (Kjellén 2000, 2006; Bayliss and Tukai 2011; Smiley 2013). In each case, local and regional contexts and nuanced differences in neoliberal forms of urban provisioning create different options and conditions for residents. These studies provide a foundation for our work because they provide a basis for comparison with our case study in Niamey. In Niger in particular, several factors guide water policies: the location of the piped water network, land tenure status, and the age of neighborhoods, among others (Bontianti et al. 2014; Hungerford 2012). However, these factors alone do not provide an accurate picture of water access and do not account for the service provided by urban water vendors.

Water vendors first emerged on the urban landscape in Niamey in the 1950s (Hungerford 2012), and their methods of water delivery and importance within the water network have continued to evolve. Residents of Niamey who we interviewed confirmed this. According to one Tuareg from Mali who has lived in Niamey since the 1960s:

> The *ga'ruwa* history dates back to my knowledge since the Diori era [Hamani Diori served as the first president of the independent Republic of Niger, from 1960 to 1974]. Actually, during that era, pumps were rare in the city of Niamey.… At that time, the *ga'ruwa* carried the water on

their heads while saying, "*Ga garwar ruwa guda*" [Look at me carrying one kerosene can full of water]. It's beyond that; it's gone. This activity experienced a big advancement. Today the *ga'ruwa* have carts and plastic containers.

Cheiffou Idrissa, a social anthropologist in Niamey, more specifically describes the material changes of *ga'ruwa* water delivery:

> In the past, *ga'ruwa* sold water using oil drums called *tukku* in Hausa (white metal containers for product preservation such as petroleum). Two oil drums fastened to a pole, together is called *talla* in Hausa. Today the oil drums are disappearing and being replaced by 20- and 25-liter cooking oil containers.

The transition from *tukku* to plastic cooking oil containers occurred in the late 1980s and early 1990s, in part due to health concerns stemming from the reuse of containers used to store and transport petroleum. Today, some Niameyans express concerns about plastic leeching into the water—from both the transportation containers used by *ga'ruwa* and the large plastic containers used for home storage (Hungerford 2012). These large, plastic storage containers, imported from China, are replacing traditional *tuluna* made by Zarma women of the village of Boubon, which is situated about twenty-five kilometers outside of the city.

By about 1990 Tuareg and Fulani men dominated this occupation and have ever since. (In Niamey, brickmaking is the other job dominated by Tuareg migrants.) Mobile water selling is an entry-level job for these migrants, though some remain in this position for many years. Tuareg and Fulani who take the position of mobile water selling have come to Niamey due to a range of motivations and have been there for wide-ranging lengths of time. They are typically men who either have lost their herds or are trying to rebuild their diminished herds by earning cash and have been in Niamey for five to fifteen years. A few have been water carriers in Niamey for thirty years or more. Finally, some are refugees from the recent conflicts in Mali and have been in Niamey for only a few months or years. Most *ga'ruwa* live very frugally in order to save money for future investments or to send remittances to family in Mali and other parts of Niger. For example, many choose to sleep on mats on roadsides to avoid paying rent, which would consume about one-half of their monthly incomes.

The length of time needed for Tuareg and Fulani men to establish themselves as independent *ga'ruwa* varies from a few months to a few years, depending on individuals' prior connections and resources. They must integrate themselves into networks of *ga'ruwa* who are already established in this occupation and work in informal apprenticeships. They typically begin by filling in for other *ga'ruwa* who leave for visits to their hometowns. Before they set out on their own, they must have their own pushcarts and water containers. New pushcarts cost 50,000 francs (FCFA; US$100) each, and used ones sell for 35,000 francs (US$70) each. The water containers cost 750–1000 FCFA each (US$1.50–US$2.00). The pushcarts and containers are made by local blacksmiths and sold primarily at the Katako Market, Niamey's main construction materials market. Those who cannot afford to purchase the tools of the trade can rent them in order to get started. Most *ga'ruwa* learn basic Hausa and/or Zarma, the most commonly spoken languages in Niamey, to accelerate the apprenticeship process. Finally, vendors must purchase annual licenses of 4,500 FCFA (US$9) from the city to operate. From there, the job requires maintaining and developing clientele—and hard work.

Most Tuareg and Fulani men would not choose to work as *ga'ruwa* if they had other good options. Most hope to earn enough money within a few years either to return home or to open boutiques in Niamey. Nevertheless, they take great pride in their hard work to provide a crucial service in the city. Indeed, hard work is central to their identities in Niamey. Many residents of Niamey, including people of traditionally sedentary groups, recognize and respect their hard work.

Water and Ga'ruwa *Identity*

As a life and death matter throughout the world, "water plays a vital part in the construction of identity at an individual, local, and national level" (Strang 2004: 5)—similar to the ways food cultures can be important in identity construction. Tuareg and Fulani experience the life and death power of water in very direct, immediate ways, as their traditional pastoral nomadic adaptation involves daily struggles to access sufficient drinking water for people and livestock at widely scattered wells, boreholes, small holes, rivers, and seasonal ponds. In Sahelian-Saharan cultures of Mali and Niger, "if someone shows up at your tent or encampment, it is your obligation to provide [water]" (de Villiers and Hirtle 2002: 120).

In addition to the popular, short Tuareg proverb *Aman iman* (Water is life), an important Tuareg proverb explains, "Where water flows freely, people live in affluence." Tuareg value the life-giving power of water throughout the lifecycle, as explained by Walentowitz (2011: 90):

> Among the Azawagh Tuareg, a preterm newborn is bathed in water containing the *aggar* fruit of an acacia tree.... It is immersed daily in the water for a period as long as it would normally have stayed in his mother's womb. This generative power of *aggar* water is also found in the Tuareg proverb "the person emerged from *aggar* water," meaning that the person looks well and has recovered from a difficult period or a longer disease.

Water is imbued with the supernatural in Tuareg cosmology, which recognizes it as one of the four elements, along with heaven, earth, and wind (Harding 1996: 3). Dangerous spirits live in water—in particular, powerful *marabout* can regulate rain (Rasmussen 1992)—and though "normal cycles of rain and drought are accepted, extreme drought is given a supernatural explanation such as God punishing people for their sins" (Rasmussen 1998: 464). Rasmussen (1992: 109) further explains the dichotomous nature of water:

> Among Tuareg, *al baraka* is a mystic power of magical properties. *Al baraka* is found in water, since water is the means of purification, and it cures illnesses and other afflictions caused by the spirits. But water and all places which contain it are also believed to be haunted by *djinn*. Such places are therefore to be avoided, or at least prophylactic measures should be taken against them.

Nomadic Fulani endure lifelong battles to stave off dehydration in the difficult conditions of the Sahel. Women, who bear most of the responsibility for securing drinking water for human consumption, face the difficult, daily drudgery of hauling water primarily from hand-dug wells and ponds to their households (Loftsdóttir 2008: 85). Nevertheless, and similar to the lives of rural Hausa and Zarma, Fulani women create crucial female spaces at wells, "representing a significant part of their social life ... a community of women is formed around those who draw water from the same wells" (Regis 2003: 125). They share news, jokes, and gossip while helping each other in the strenuous labor. However, this companionship functions not only to alleviate physical burdens but also to provide them with protection against the dangerous spirits that inhabit water (Regis 2003: 126). Although some water spirits can "bring them fortune ... they may also prevent them from having human families and children" by "marrying" women and killing their human rivals (Regis 2003: 129).

Just as water is fluid by nature, so are Tuareg and Fulani identities and relations to water. The decline of nomadic pastoralism is largely—thought not exclusively—due to drought and decline in water availability. Tuareg and Fulani take control of their lives by becoming *ga'ruwa* and taking some control of water. Water delivery has become integral to their identities in ways that reflect on their rural past, but are constructed in new ways in the city.

Tuareg and Fulani *ga'ruwa* regard themselves as nomads who are temporarily in the city and take great pride in their abilities to combine a strong work ethic with sociocultural competence to offer important services to their families and the residents of Niamey as dignified, moral men. Their work, as some put it, adds new, modern meaning to the Tamachek expression *Aman iman* (Water is life).

Tuareg and Fulani *ga'ruwa* typically retain a strong nomadic identity even when they have been stuck in Niamey for many years. They speak longingly of their herds and their desire to return to the bush, and they maintain regular contact with their nomadic families and communities through cell phones and remittances. *Ga'ruwa* see their hard work as honorable and in contrast to lazy "white" Tuareg, who only gain their wealth through smuggling, and idle Hausa and Zarma, who while away their days in street corner conversation groups telling jokes. They also value that their strength and endurance must be augmented with sociocultural competence since their job requires the navigation of ethnolinguistic and gender boundaries. That is, they work with and for people across the ethnolinguistic spectrum in Niamey, and they must earn trust in order to be allowed into women's spaces in homes to deliver water—a matter that we will return to later in this chapter. Black Tuareg attribute all of these qualities to their racial identity and their perception of themselves as righteous Muslims.

Working as a Ga'ruwa

Ga'ruwa in Niamey typically work from sunrise to sunset seven days a week. They operate in almost every neighborhood in Niamey. Only in a few pockets within wealthy neighborhoods are they rare or nonexistent. At sunrise (or earlier during the hot season, when water does not always regularly flow through the piped system), *ga'ruwa* bring their carts to the single standpipe from which they work and fill the ten to fourteen 25-liter containers of water on their wheeled carts. *Ga'ruwa* always use the same standpipe to fill their containers for two primary reasons. First, customers are charged for delivery based on the distance the *ga'ruwa* has to travel from the standpipe to the customer's home. Using the same standpipe to obtain water means that the cost to the consumer is consistent. Second, there is an informal agreement between *ga'ruwa* that each one will use the same standpipe every day so as not to infringe upon the territory or customers of other *ga'ruwa*. *Ga'ruwa* are not assigned to particular standpipes but become affiliated with certain standpipes through the integration and apprenticeship process mentioned earlier. Using multiple standpipes to fill water containers would be a violation of this unwritten agreement.

Delivering water is physically strenuous. *Ga'ruwa* push their metal carts with full water containers on mostly unpaved, uneven neighborhood streets. Upon reaching a customer, the *ga'ruwa* carries the containers of water, two at a time, into the customer's home or compound and empties the water into larger storage containers (plastic barrels or clay pots) indicated by the customer. The *ga'ruwa* typically must return to the standpipe to refill their containers after delivering water to three to five customers. One *ga'ruwa* described the physical impact of the job: "At nightfall, my entire body hurts, especially my chest and hips. There are some places where there is too much sand, stones, and inclines, where we must push [the carts] with a lot of energy." According to another *ga'ruwa*:

> Today, if you find me another job, for example, security, I will give up this work that I've done since the time of Ali Chaibou [1987]. You see that makes me old. At that time, there were no carts with containers, only *le tagala* [two buckets attached to a pole and carried on the shoulders].

Ga'ruwa also suffer with painful cracked feet due to standing in pools of water around standpipes. Almost all *ga'ruwa* wear closed toe, fully plastic shoes of the same brand made in China to combat this problem. As one *ga'ruwa* explained:

> It is to protect us against the water. These shoes do not weigh much, they aren't too heavy to wear, they adapt to all places and all circumstances, and they are like sports shoes. If we bend over to push the cart in the sand, they do not break. Their price varies from 1,100 to 1,250 FCFA.

If business is good, then the *ga'ruwa* take occasional breaks, usually in a shaded area close to their standpipe. In the hot season, these breaks last longer if they are fatigued or if little or no water is flowing through the standpipe. Understanding the water needs of their customers, the *ga'ruwa* try to deliver water around the same time of day every day (morning, midday, afternoon). The regular delivery schedule ensures a consistent customer base.

Ga'ruwa face other challenges in their work, particularly due to dangerous situations on the street and stubborn customers. One 35-year-old Nigérien *ga'ruwa* from the Kandaji-Tillabery region (near Mali), described these challenges:

> If you look in our carts, there is a staff hung on the side, that is for approaching certain situations (dog attacks, or bandits, or someone

refusing to pay us). The second problem that we have is the refusal of certain people to pay us at the end of a predetermined consumption period. They tell us to go complain everywhere we want. When it comes to us, we leave them with their conscience, as even Islam spoke of the value of water in the life of an individual. We are not like the SEEN, who, when you do not pay your bill at the end of the month, cut you off. These are problems that we regularly encounter in our work that will push us one day to establish a union to defend our interests.

Although none of the *ga'ruwa* that we interviewed mentioned the difficulty of doing their jobs during Ramadan, we sense that this is a particularly challenging time for them, especially when Ramadan falls during the hot season. Their customers consume almost as much water daily during Ramadan as they do daily during the rest of the year, because from sunset to sunrise they cook, drink lots of water, and bathe. *Ga'ruwa* must continue their daily water deliveries. They must push heavy carts full of water and lift heavy containers full of water all day, but they cannot drink any of it.

Ga'ruwa straddle the formal and informal economies of Niamey. As mentioned earlier, to work as *ga'ruwa* in Niamey, they must acquire annual licenses from the city at a cost of 4,500 FCFA (US$9). This amount represents a 1,000 FCFA (US$2) increase between 2013 and 2014, a condition that frustrated many *ga'ruwa* in our study. Other than paying the annual fee, no other official requirements for *ga'ruwa* exist. The *ga'ruwa* do not pay taxes or report income to the city. Thus, after paying the annual fee, they shift to an economic activity (water delivery) that falls into the informal economy, although they still use formal structures—the piped water network—to perform this task. For this reason, Bakker's (2010) definition of "hybrid economies" applies to this form of water vending.

In Niamey, neighborhood men manage standpipes. These men are of various ethnic groups, including Hausa, Zarma, and Songhay, but they are disproportionately Tuareg and Fulani. In many neighborhoods, this job rotates among a few people. Standpipe managers have three primary responsibilities. First, they collect money from *ga'ruwa* for the water they take from the standpipes to fill their containers. These managers also collect money from individuals who live close enough to the standpipe to transport water themselves. The managers, in turn, pay the city (and by extension, Veolia) for the water use. Second, they must maintain the cleanliness of the pipes and the areas around them; otherwise they face fines imposed by city hall, which regularly sends inspectors out to check on the condition of standpipes. Third, standpipe managers mediate and resolve disputes between their customers.

Customers are expected to respect rules of decorum. *Ga'ruwa* are the primary customers, and they should park their carts by order of arrival and patiently take turns. However, when a customer arrives with a single bucket, the *ga'ruwa* should allow him to fill it if the *ga'ruwa* is at the beginning of his loading. *Ga'ruwa* who have already filled two or more containers expect to complete filling all ten or twelve containers before relinquishing the hose to a person with only one container.

During our fieldwork, we witnessed several disputes precipitated by *ga'ruwa*'s reluctance to surrender water hoses to non-*ga'ruwa*. In one typical scenario, a customer with a single bucket politely attempted to jump to the front of the queue. An angry *ga'ruwa* shouted, "Didn't you see me here! You must hold on; I am not obligated to give you the hose; be respectful of people." The local resident aggressively answered, "What is this? Help me! Why do I have to wait for you to fill all of your containers when I only have one container?" As the situation appeared to be escalating to a potential fist fight, the manager intervened to restore peace, "Hey you two, that is not how it is done. You must respect people; things can be negotiated; do not use force. *Ga'ruwa*, forgive him; have patience." The *ga'ruwa* accepted this mediation and handed over the hose.

Ga'ruwa pay the standpipe managers 15 FCFA (US$.03) per 25-liter container of water filled, and the *ga'ruwa* collect 25–50 FCFA (US$.05–.10) for each 25-liter container of water delivered, depending on the distance between the customer and the standpipe. *Ga'ruwa* also make spontaneous sales of water on the street, for example, to commercial truck drivers about to embark on long trips. Thus, a *ga'ruwa* who pushes a cart that carries ten containers earns a profit of 100–350 FCFA (US$.20–.70) per cartload, after paying the standpipe manager for the water. Drawing from our observations while shadowing *ga'ruwa* on their daily rounds, refilling containers takes about ten minutes when water is regularly running to public standpipes, delivery trips last about forty-five minutes each, and *ga'ruwa* make about twelve delivery trips daily. Hence, *ga'ruwa* earn 1,200–4,200 FCFA (US$2.40–8.40) daily. Although this income is small, it is above average and fairly consistent, as most households require daily water delivery. According to one *ga'ruwa*:

> It has been three years that I have been working as a *ga'ruwa* in Niamey. Thank God. I have found fulfillment here; in fact I got married here. I am buying some fertilizer to send to the village to put in my garden. I sometimes send money to my parents who are in the village. Anyway, I make a living here.

Ga'ruwa encounter many circumstances that can lead to loss of income. For example, they earn no income on days they do not work for whatever reason and on days during the hot season when water may slow to a trickle or cease flowing altogether to public standpipes, especially those most distant from water towers. In addition, in recent years SEEN has offered promotional discounted rates to install pipes in homes and on the first month's water bill, leading many to switch to piped water. For example, Issoufou, a former *ga'ruwa* customer explained, "About a month ago we stopped using the services of *ga'ruwa*. In fact, the homeowner recently installed a pump."

Several of the *ga'ruwa* that we interviewed complained about a drastic drop in customers. As one *ga'ruwa* put it, "Now we don't have many customers; many people have access to their own pumps at home." However, we think that this downturn in business is localized. We do not think that SEEN can switch enough households to piped water to offset Niamey's continuing growth. Niamey is expected to remain one of the world's fastest growing cities, with annual growth rates projected at 5 percent for the next fifteen years (Demographia 2015). Much of Niamey's population growth is occurring in periurban areas that are not yet connected to the piped network. These conditions will continue to provide ample opportunities for *ga'ruwa* to find work. Furthermore, Niamey's ever-growing low-income populations cannot do without the services of the *ga'ruwa*.

Water delivery is so essential in Niamey, and such an important element of daily life, that cultural landscapes have been created by the process of water delivery and can be found in almost all neighborhoods. For example, clusters of *ga'ruwa* carts with yellow containers often identify locations of public standpipes (especially for outsiders like us), rather than the standpipes themselves, which tend to visibly blend into the surrounding landscape, particularly those located along busy streets (Figure 4.2).

The *ga'ruwa* are part of the economic landscape in Niamey. They travel between busy intersections, where public standpipes are often located, and small, quiet, neighborhood streets. The metal carts and yellow containers common to all *ga'ruwa* specifically identify them to passersby as water deliverers. However, *ga'ruwa* try to personalize this common image by adding decorations to their metal carts, such as flags of their country or region of origin, colored ribbons, or plastic flowers (Figure 4.3). Not only does this personalization mark contrasts between *ga'ruwa* and the public image with which they are associated, it also serves a practical purpose, as it distinguishes one metal cart from another, an essential characteristic around busy public standpipes.

Figure 4.2 A cluster of *ga'ruwa* carts around a public standpipe in a peripheral neighborhood of Niamey. Photo by the authors.

Solidarity among Ga'ruwa

Straddling the formal and informal economies of Niamey, *ga'ruwa* cooperate using a union-like form of solidarity. This sets them apart from water vendors elsewhere in Africa who compete with each other for customers and water sources (Cairncross and Kinnear 1991; Kjellén 2000, 2006; Smiley 2013; Whittington et al. 1989).

Informal economy workers across Africa have found it difficult—but not impossible—to establish labor unions (Lindell 2010; Rizzo 2013). No laborers in the informal sector in Niger are unionized. The crucial reason for this is that informal economy workers are not involved in direct employer-employee relationships. This includes the *ga'ruwa*, who do not work for employers in the standard sense but rather for dozens of individual clients. Drawing from Rizzo's (2013) analysis of the successful unionization efforts of informal bus drivers in Tanzania, the *ga'ruwa* hold "structural power" because about two-thirds of Niameyans rely on their daily services. The *ga'ruwa*, however, lack "associational power" to this point, such as links with trade unions that typically consist of state employees—often key to founding unions among informal economy workers. *Ga'ruwa* have not found it necessary

Figure 4.3 A *ga'ruwa* cart with decorations. Photo by the authors.

to push for institutionalized unions due to their effective informal relationships and because there is not an oversupply of *ga'ruwa*.

Informal *ga'ruwa* associations in Niamey typically include a core group of men who consider themselves family, broadly defined. That is, they may not be brothers or even cousins, but they may be distantly related and are usually from the same hometown or region of Mali. Furthermore, these connections are about as likely to draw on patrilineal as matrilineal ties, reflecting changing kinship systems across many Tuareg subgroups.

Understanding the solidarity among and between Tuareg and Fulani *ga'ruwa* requires some historical perspective on their relations. In the pastoral zones of Niger and Mali, Tuareg lineages and Fulani lineages have their own inherited wells. They use each other's wells only with permission, which is rare, because it involves moving large numbers of animals through potential pasturage. However, in some cases they have learned to cooperate and share state-sponsored boreholes, some of which have been in place since the early colonial period a century ago (Wilson-Fall 2015). Furthermore, Tuareg and Fulani share some cultural affinity, particularly through their identities as herders and nomads. Today, several prominent musical performance groups in

Niger—such as Etran Finatawa—include both Tuareg and Fulani. This urban, context-dependent pastoral nomadic identity in segmentary opposition to a foreign and hostile sedentary world is quite remarkable given the longstanding conflicts over access to pasture and wells between Tuareg and Fulani in Niger—Mali border areas that have become increasingly lethal in recent years.

The solidarity among and between Tuareg and Fulani *ga'ruwa* in Niamey is crucial to their success. They often work together or at least agree not to infringe on each other's pre-established territories and clientele. They fill in for fellow *ga'ruwa* when they cannot work due to illness or other reasons. They share information, allowing them, for example, to collectively boycott households that consistently ask for credit but do not make good on it. The city grants all *ga'ruwa* license requests, as the fee provides income for the city, so theoretically there is no limit to the number of *ga'ruwa* that operate annually in Niamey. Thus, the informal solidarity between *ga'ruwa* helps to maintain order and ensure customers' needs are met. Water delivery is not a free-for-all in Niamey.

The solidarity between *ga'ruwa* is a reality recognized by everyone— *ga'ruwa* and their customers. For example, a *ga'ruwa* at a tap stand explained:

> That is true, we do stick together, if you have a problem with one of our comrades, we will not serve you because we tell ourselves that we can have the same problems with you as well. This solidarity is explained by the fact that we do the same work. If I sell water to one of my comrade's clients while he is away, without his authorization, it is hypocrisy, betrayal. When a *ga'ruwa* sells water to another's client right in front of him and the latter *ga'ruwa* doesn't react immediately, be sure that they will settle it once they go back to their base to rest.

Similarly, Assoumane, a tea vendor who relies on *ga'ruwa* for his water supply, observed:

> They have solidarity.... You have your *ga'ruwa* who brings you water every day, and if there is a day where he isn't there and you ask another to bring you water, he will never do it; he will tell you to wait until your *ga'ruwa* comes back. No matter how much you shout, he will not look at you, even if he has extra water in the containers; he prefers to return to the tap stand to refill his cart. Personally, my *ga'ruwa* brings me water in the morning, but sometimes I need water at other times; at those times, I must negotiate with their elder in order for them to serve me in the evening if I need it. He does this service for me because he has been with us for a long time. It's at my house that he eats breakfast; he is from our ethnic group. The last time that my *ga'ruwa* wanted to return to the country, he came to introduce me to his replacement.

According to Abdoul Malik, another *ga'ruwa* customer nearby:

> Yes, that is a reality. In fact, one day when Mamou, the owner of the tap stand that you see in front of us, had a problem with a *ga'ruwa*, all of his *ga'ruwa* clients boycotted his tap stand to join the neighboring tap stand. They truly do stick together. This would be explained by the fact that, in general, they come from the same family or the same locality.

Moussa Mahamadou, another customer, explained:

> They do stick together. One must avoid piling up outstanding debts with a *ga'ruwa*: as soon as he informs his friends, they as well will avoid you. I remember well the time when one was subscribed to a *ga'ruwa*: if he traveled, he would be replaced by someone. Before leaving, he would consult his clients and offer them the position. It's absolutely normal that they stick together since they practice the same profession. It's like trade unions for workers.

Despite the comradery in the context of work, however, Tuareg and Fulani *ga'ruwa* tend to socialize in ethnically segregated groups after hours. Most *ga'ruwa* are either single or do not have their wives with them in Niamey. As a result, they cook and eat together and often live together—even if only sleeping side by side outdoors on mats. As Abdoul Malik, a customer, observed:

> The concrete example is that of the local group of *ga'ruwa* that are composed of ten or so elderly people and young people. They contribute enough between them to pay for the essentials. This does not prevent those who have some money to pay for meat or other things to accentuate the courses. The young ones prepare the dishes in turns; as for the old ones, they are exempted from cooking. Between them, they tease each other a lot, provoke each other, and fight; each time that I observe them, they make me understand that it is between them, that they are parents, and notably, cousins.

In other words, the roles assigned in nonwork situations are a result of ethnic and familial relations, rather than of their role as *ga'ruwa*. However, it is the shared experience of delivering water in particular cultural contexts that results in a comradery that crosses ethnic lines.

Gender, Ethnicity, and Water in the Sahel

A nuanced understanding of the culture of *ga'ruwa* and their role in the urban water regime of Niamey requires an appreciation of the symbolic value of water and the social relationships that revolve around it, particularly gender and ethnic relations. Furthermore, exploring the relationships between *ga'ruwa*, water, and ideas about gender and ethnicity helps us answer our original questions about why water vendors in Niamey are mostly male Malian immigrants.

In Niger, the procurement and delivery of water is an activity that is gendered in particular ways among different ethnic groups. Among the sedentary agriculturalist Hausa, Zarma, and Songhay, who live in rural areas, village women gather and deliver water as mothers and wives—one or two 25-liter containers at a time, carried either atop their heads or on either ends of poles balanced over their shoulders. This is regarded as "women's work" and women are not paid for this service. Furthermore, women are expected to serve water to men, never the reverse. Although this is difficult and time-consuming labor, they have carved out one of few public women's spaces at village wells. Pausing at wells offers one of the few moments in women's busy days to gather solely in the company of women to share news, tell stories, and laugh.

In contrast, among urban Hausa, Zarma, and Songhay in Niamey, only men gather and deliver water, and they do so for pay (women also serve water to men in the city). Urban men in Niamey claim that it is a good thing that urban women are "relieved of this difficult chore," but we note that this gain comes with the loss of an important women's space and important potential, paying jobs. Furthermore, while women do not gather and deliver water, they are responsible for cooking, washing, cleaning, and bathing children, all of which involve water. Women are also responsible for storing water for short-term use, either in traditional *tuluna* or in cheap imported plastic containers. Thus, in the urban context of Niamey, women are responsible for using water (and, as customers, for securing its presence in the household), but they are removed from the delivery process.

Among the traditionally nomadic Fulani and Tuareg, the procurement and delivery of water is also gendered but in different and less dichotomous ways than among sedentary peoples. Among the Fulani, men and women are far less segregated than their sedentary neighbors. Men typically draw water for livestock, especially cattle, whereas women gather water for household use. However, Fulani women help in getting water for animals where there are not enough men to do it. They may specialize in distributing it to goats, sheep, and calves, depending on the situation. Women also herd, especially young girls, if there are no males to do so (Wilson-Fall, personal communication).

Since independence, the nomadic pastoral way of life among the Fulani has become increasingly precarious, as many have lost their herds and have had no choice but to take up farming or urban labor.

As noted by Kristín Loftsdóttir in her long-term study of WoDaaBe (a Fulani subgroup in Niger), important scholars of the Fulani such as Dupire have, for more than fifty years, documented "an element of shame associated with being engaged in occupations other than herding one's own animals" (Dupire 1962: 126–27 cited in Loftsdóttir 2008: 154).

The passage of time has brought the realization that diversified economies drawing on rural and urban, nomadic and sedentary ways of making a living are necessary, and hence the shame of doing work outside of the pastoral economy has diminished. Loftsdóttir (2008: 154) "did not find WoDaaBe migrant workers being generally ashamed of their work in the city. Some individuals I interacted with were on the contrary relatively proud of their work in the city, emphasizing their importance for providing a security net for those in the bush." Loftsdóttir (2008: 155) goes on to explain, "Shame seems thus today to be more connected to failure of earning income from one's activities in the city rather than working as a migrant laborer [per se]. Some occupations, such as tea selling, rope making, and water carrying are [generally] associated with failure, and thus shameful." In contrast, making and selling handicrafts and art is generally more lucrative than other typical jobs Fulani get in the city, and it is not considered shameful. The Fulani *ga'ruwa* that we interviewed in Niamey did not express shame or embarrassment in doing this work, though several told us that they would much prefer to be herding cattle in the bush. There are several possible explanations for this. For example, they may have felt too proud to share this feeling with strangers, or we may simply have encountered men who were earning enough money to send some home regularly.

Among the Tuareg, the situation is even more complicated, in part because they are largely matrilineal, as there are regional, rural-urban, nomadic-sedentary, and class-related differences (Rasmussen 2009). In rural Air in Niger, in seminomadic and sedentary villages, both sexes draw water from wells; usually, men nowadays get it from wells and a few motor pumps to irrigate their oasis gardens, and young boys tend to draw it from wells to water livestock herds. Women and girls draw water from wells for household use and transport it physically or on donkey carts if they are relatively close to home. Men use camels to transport water for domestic use where wells are far from home. If possible, adults prefer to have their adolescent children do these arduous tasks, and if they can afford it, some families hire men to draw it.

In rural northern Mali in the Adragh and Kidal region, in seminomadic villages, most Tuareg women of nonservile descent, who avoid most physical labor, disdain this task and hire Bella men of servile descent to do this. In Saharan multiethnic towns such as Agadez and Kidal, the situation is again different. Tuareg families there tend to hire others to do this work if they can afford it. In Agadez, Niger, most *ga'ruwa* are Hausa men, particularly labor migrants from western Niger villages that have suffered from droughts, locust invasions, and unemployment. In Kidal, Mali, Tuareg families, as in Agadez, tend to hire various non-Tuareg to fetch water for their households. As in Agadez, most Tuareg in Kidal try to avoid this task and many hire other non-Tuareg Malians to do it, as well as other domestic tasks, often Dogon men to fetch water and Dogon women to help cook. In sum, there are class and ethnic differences that complicate the gender distinctions, but in urban settings in Tuareg regions, non-Tuareg men and a few Buzu or Bella tend to draw water for households who can afford to hire them.

Returning to Niamey, we can see that the job of delivering water places the *ga'ruwa* in a gray area in terms of gender relations. *Ga'ruwa* deliver water to households and compounds that cannot afford direct connections to the piped network (Youngstedt et al. 2016). Not surprisingly, then, *ga'ruwa* are most active in poorer neighborhoods, particularly those in Communes III, IV, and V, and in some newer neighborhoods on the urban-rural fringe in areas where the piped network has not yet expanded. Most neighborhoods in Niamey are class segregated but ethnically integrated, especially considering that ideas about ethnicity are rather fluid in Niger (Youngstedt 2013). People from the same village or region of Niger often live close to each other within neighborhoods. Thus, *ga'ruwa* provide a service to a socioeconomic class that is excluded from the piped network, and this includes people of all ethnic groups living in Niamey.

During the day, male family members are typically absent from households and compounds in Niamey, usually because they are working, trying to find work, or visiting friends and family. In this situation, it is highly unusual for male nonfamily members to be allowed into households or compounds. However, the *ga'ruwa* are an exception. Not only do *ga'ruwa* enter compounds, they carry water into typically female spaces, such as the cooking or washing areas, to empty the water containers into larger storage containers. Assumptions about water perhaps explain this acceptance. Among sedentary agriculturalists in rural Niger, obtaining water and transporting it to households is a job that is considered women's work. Although men in Niamey deliver water, it is perhaps the activity, rather than their gender, that makes their presence in the home acceptable, even when male family members are absent.

The informal agreements regarding professional practice also explain the exception made for *ga'ruwa*. Because *ga'ruwa* have regular customers, they are not strangers to the people in households and businesses where they deliver water. *Ga'ruwa*, instead, regularly appear at the door to the compound every day, or every other day, to deliver water. This is also why, when a *ga'ruwa* temporarily turns his clientele over to another, such as in anticipation of a long trip, he introduces the replacement *ga'ruwa* to his clientele before his departure.

Conclusions

Tuareg and Fulani are adapting to crises through diversified economies, structural fluidity, and changing and increasingly flexible kinship systems. Tuareg and Fulani *ga'ruwa* are integral contributors to transnational communities involving both other urban labor migrants and rural nomadic and seminomadic pastoralists. Their identities are linked to pride in their ability to support their families and rural communities through remittances earned by their hard work. Indeed, few pastoral families could survive without the assistance of *ga'ruwa* and other urban labor migrants (Giuffrida 2010: 26). Many hope to return to their rural communities once they have the resources or there is security, whereas others seek to parlay their *ga'ruwa* earnings to invest in dry goods boutiques, enhancing their capacity to contribute to their scattered communities.

We now return to our key original question and its corollary: Why are *ga'ruwa* jobs in Niamey dominated by immigrant Tuareg and Fulani men from Mali? Why do Nigérien men in Niamey, particularly from the traditionally sedentary Hausa, Zarma, and Songhay, avoid this job? The answers are intertwined.

Nigérien men avoid this work in Niamey due to dignity preservation and shame avoidance. Nigérien men (of settled agricultural groups) consider water delivery to be inferior "women's work" and feel that women should serve men water. Furthermore, Nigérien men feel embarrassed to perform menial, physical labor in public (under the gaze of family, friends, coethnics). Reflecting views we heard repeatedly, a Hausa standpipe caretaker in the Koubia neighborhood observed:

> The young Nigériens, in particular the city dwellers, do not want to do the work of a *ga'ruwa* because they are big-headed. They do not want to suffer as well; they are ashamed. But looking to make money by your own means is better than stealing or begging. If you see some Nigériens

doing the work of a *ga'ruwa* in Niamey, assure yourself that they come from the countryside.

Several of the participants in our study indicated that Nigérien men of traditionally sedentary farming groups are willing to do this job when they are far from home, for example as labor migrants in Nigeria and Ghana or in Northern Niger. Conversely, we learned that Tuareg and Fulani decline *ga'ruwa* work in traditional Tuareg and Fulani settlements.

Tuareg and Fulani accept this job because they have very few other options in Niamey, but it is also a conscious choice—the role of handling water is important in the sedentarization processes that form their urban identities and provides opportunities to accumulate wealth and return to the pastoral life if they wish. Their sense of self (and the perception others have of them) as pious, trustworthy, and hardworking can be articulated through this work of water delivery. Fulani and Tuareg water carriers are recognized as refugees from a bad situation. In Niamey and other towns of southern Niger, they are strangers in a way that the Hausa and Zarma are not (Shack and Skinner 1979). They, like Hausa or Zarma migrants in Ghana and elsewhere in West Africa, are far from home.

These developments in the nature of water access and delivery in Niamey are ironic in two ways. First, a century ago, nomadic Tuareg and Fulani dominated sedentary Hausa, Zarma, and Songhay. Now they have to do what the latter look down upon as inferior "women's work." While living a sedentary lifestyle that they consider inferior to nomadism, Tuareg and Fulani men are key players in the most critical form of mobile water access in the city. Second, many Tuareg and Fulani have lost herds due to lack of water (and other factors, including state policy) but now must work with water.

The *ga'ruwa* play an essential role in urban water access, operating under a global neoliberal capitalist system using contemporary technology. At the same time, the process of accessing water in Niamey is steeped in cultural traditions and values that vary by ethnicity, gender, and class. The job of water delivery in urban contexts has created distinct identities among Tuareg and Fulani *ga'ruwa* that they feel distinguish them as a group from other ethnic groups in Niamey. It is both the *loss* of water (that compromised their herds and nomadic lifestyle) and the *provision* of water that make their work with this commodity an important element of their urban cultural identities. The *ga'ruwa* are not helpless—they demonstrate agency within a complex system of water commoditization—and the trust and income they are able to gain through their work is important to their identities. These identities play out in the struggle for everyday water access in Niamey—a struggle linked to the commoditization of water and its delivery within a global, capitalist system during a time of drought and desertification in a Muslim city and a struggle that has created new symbolic meanings for those involved in its consumption and delivery.

Note

1. Parts of this chapter were first published in *African Studies Quarterly* 16(2): 27–46, 2016, under the title "Water Vendors in Niamey: Considering the Economic and Symbolic Nature of Water." The original article was coauthored by Mr. Cheiffou Idrissa.

5

"Pure Water" in Niamey

When we began studying the sachet water economy in Niamey, one of our biggest challenges was securing interviews with the youths who walk the streets of Niamey selling the cold one-half-liter plastic bags from coolers secured to pushcarts or from bowls carried atop their heads. Adults in Niger typically do not engage with young people to inquire about their work and the other kinds of questions we needed to ask, and young people would be understandably suspicious of two foreign, white adults who approached them with questions about *piya wata*. So, we once again asked our loyal gatekeeper, Hassane, if he might introduce us to some of the young people selling sachets on the street and help explain to them why we had "strange" questions but that we were only curious.

We started our first day of interviews on a side street close to the Grand Marché. Not surprisingly, this large market is a popular location for sachet water vendors because it attracts large numbers of people from all over the city. It did not take long to encounter a group of sellers, and Hassane patiently explained to them who we were and that we had a few questions about *piya wata*. We waited off to the side. Soon, two brave sellers came over to us; we introduced ourselves while Hassane translated (these two did not speak Hausa), and they agreed to answer our questions.

A white couple on a crowded side street by this large, congested market attracts some attention. Because sachet water sellers travel in groups, others noticed us talking to the original two sellers and came over to investigate. We were soon surrounded by sellers, some watching quietly, others eager to answer our questions. Hassane's presence helped legitimize our position. If we were with a Nigérien man who is well respected in the neighborhood, then we were trustworthy.

Young people—almost entirely boys—play a key role in Niamey's sachet water economy. They walk the streets of Niamey selling one-half-liter bags of water from coolers on pushcarts and from bowls perched atop their heads. (Adults rarely sell sachet water on the street, though they—including a few women—do sell it in small shops.) Children and youths vending water on city roads typically walk together in groups of two to four. We were curious about this practice since it would seem to put them in direct competition with each other, so we asked them about it. They explained that they work together primarily to minimize boredom. Our interviews revealed a more interesting practice. Virtually all of the children and youths with whom we spoke indicated that they are working for women— their mothers, grandmothers, and neighbors. Women of Niamey are almost entirely absent from the public view in the city's water economies. Only men work as *ga'ruwa* and standpipe managers. Families that are too poor to afford *ga'ruwa* services send boys to fetch water from standpipes. We realized that we needed to meet some of these women.

Mariama, a Zarma woman in her early thirties, is the most experienced and successful behind-the-scenes sachet water vendor that we encountered. She began selling sachet water in unlabeled hand-tied bags fifteen years ago. As labeled machine-sealed bags surged into the market, she converted to selling them about twelve years ago. Mariama explained that she is motivated to help her children and family since her husband is unemployed, and she is motivated to be a sachet water vendor because she can make "easy money" from the comfort of her own home. (In theory, in Islam, husbands are responsible for providing all of life's necessities to their wives and children, and wives have the right to retain any other own earnings. In practice, Nigérien women will not let their families suffer if they can help it, and many are primary or important breadwinners.) Mariama capitalized on three key factors: her husband purchased one of the two refrigerators that she uses many years ago while he was still employed, cutting her initial capital investment in half; there is a high markup on the sachets that she buys in bulk very cheaply; and she has two children to do the work of selling them.

Mariama purchases sachet water packaged in groups of twenty bags for 150 FCFA—or 7.5 FCFA each—from a neighbor who produces the bags in his home with an automated filling and sealing machine. Her children sell each bag on the street for 25 FCFA, the standard price throughout the city. Hence, Mariama earns 17.5 FCFA profit per bag, a staggering profit ratio! She estimates that her children sell 480 bags daily, meaning that she earns 8,400 FCFA (about US$16) daily or 252,000 FCFA monthly. Mariama has expenses that cut into this profit, such as a monthly electricity bill of 15,000 FCFA, and she pays her older child

15,000 FCFA monthly and her younger child 10,000 FCFA monthly, but she earns a considerable income given that the per capita daily income in Niamey is about 500–1000 FCFA. Mariama provides food and clothing for her husband and children in Niamey and sends about half of the rest of her earnings to her family in Dogondoutchi 275 kilometers east of Niamey.

This chapter seeks to uncover the backstory of one element found in many streetscapes in Niamey: discarded half-liter plastic bags—called "sachets" that once contained cold water. Residents of Niamey refer to this water as *piya wata* (pure water), a name whose origins are explored later in this chapter. Discarded bags are the final product of commodity and value chains of water production, exchange, and consumption that fall mostly outside government-regulated water services. By working backward from the discarded bags to those who consumed the water in them, to those who sold them, and finally to those who produced them, we demonstrate that four key characteristics, mainly linked to commodity production, affect the economic value of sachet water: the label, the temperature of the water, the time of year the bag is sold, and the apparent purity of the water. All of these are influenced by the social context in which they take place. We explore these elements of economic value alongside cultural context and social relationships to uncover a complex political economy of water selling with implications for the environment, poverty and socioeconomic status, gender, local economies, and global trade.

Commodity Chains and Value Chains

The material culture of plastic water bags cannot be understood without considering the social and cultural processes that create them (Buchli 2002). Plastic water bags move through the commodity chain while individuals attach meaning and value to them. In this way, water bags have a rather fluid existence (Buchli 2002) or "social life" (Wagner 2013; see also Orlove and Caton 2010; Bridge and Smith 2003), not only because of what the bags contain but also because the bags and their contents pass through the phases of extraction, production, exchange, and disposal. The material items themselves have economic and symbolic value, although that value changes as the bags move from large rolls of plastic sold in markets, to containers holding cold water, to things that are no longer needed once the water has been consumed. (The fact that the bags were commodities sold to people three times is just one part of the backstory, the commodity phase.) The stories surrounding these phases must be considered in any analysis (Page 2005). Thus, something that has been discarded, and is now part of what we call a landscape of waste, still has considerable value—though of a different sort—particularly in the way it communicates cultural values and practices and is connected to larger stories and systems.

As plastic water bags move through the commodity chain, their value changes. The global value chain is as important as the global commodity chain in understanding the story behind the final, discarded bags in the landscape. Global value chains imply not just a connection between stages through which a commodity passes (and the actors involved) but also a connection between commodities, activities, social relationships, and perceptions entailed in each link of that chain (Cook et al. 2004; Kaplinsky and Morris 2001; Ricketts et al. 2014; Tsing 2013; Zylberberg 2013). However, value is created and distributed unevenly throughout the commodity chain, resulting in changing power relationships between actors at each stage. For example, Swyngedouw (2004: 1) contends that value is added in the production stage. He argues, "In capitalist cities … [the] circulation of water is also an integral part of the circulation of money and capital." Water—and the objects used to extract, purify, transport, store, and consume it—has passed through processes of commodification that add to its value.

In contrast, Appadurai (1986: 13) challenges this Marxian account of value production by asserting that value is assigned during the process of exchange, which occurs within cultural, political, and economic contexts. In her study of matsutake mushrooms, however, Tsing (2013) demonstrates how value is added both to the commodity itself and to the process of acquiring it at multiple phases in the commodity chain as the mushrooms are assigned different values by different actors when social relations are established and transformed. In other words, no one phase in the commodity chain is more important than others in assigning value.

Global commodity chains and value chains are a result of the movement of goods in the global economy. While many have benefited from globalization, sub-Saharan Africa's increased participation in the global economy has brought decreased income shares, especially for marginalized groups (Kaplinsky 2000). Thus, development efforts that focus on increasing the value in global value chains often emphasize the process of upgrading by increasing the efficiency of production, producing higher-value goods, or using new skills to expand into new economic sectors (Humphrey and Schmitz 2002). The process of upgrading takes many forms but usually involves some type of empowerment through which "marginalized individuals and groups are able to exercise a meaningful level of control over the progressive realization of their own well-being" (Macdonald 2007: 794).

Much of the research on commodity chains and value chains in sub-Saharan Africa focuses on case studies of cash crops and natural resource exploitation. Themes include the rise of buyer-driven global value chains (see, for example, Gibbon and Ponte 2005), increasing profits for smallholders by upgrading to high-value crops (see Zylberberg 2013), perceptions of risk in this process (see Ricketts et al. 2014), and the potential for poor countries to further integrate into, and benefit from, the global economy. Our study of sachet water is informed by these case studies but is different from them in several ways. We return to these ideas later, but first we situate sachet water production in the broader historical context of water provision and privatization in Niamey, and then we focus on each link in the commodity and value chain, beginning with the discarded bag.

Water Privatization and the Rise of Sachet Water in Niamey

An important element in the backstory of sachet drinking water in Niamey is its emergence in the city's changing water regime at a time when urban water supplies across the globe were becoming complex and fragmented. Sachet water is one example of Bakker's "hybrid economy" categorization, whereby both public and private entities are involved in providing water to the "urban unconnected," as well as those with direct connections (Bakker 2010: 6). She introduces a tripartite typology to categorize water industry actors: governments, private corporations, and community groups or individuals (Bakker 2010: 26). While all three groups are involved in the story of sachet water in Niamey, it is this third group—one made up of individuals who are neither government employees nor members of private corporations and one that serves the community, defines goals based on community needs, and considers community opinion—that is most prominent in our story.

The privatization of water in Niger is but one outcome of global corporate-driven neoliberal policies (such as structural adjustment) that have pushed for the privatization or shared management of utilities and services traditionally managed solely by the state and have promoted the shrinking of government spending on and ownership of urban utilities and the fragmentation of state management. This process of privatization occurs in many forms and, in the case of sachet water, includes individuals working for themselves, in many cases producing the sealed water bags in the privacy of their own homes and either selling water in public spaces or in private, locally owned shops. We estimate that the cottage industry accounts for at least 90 percent of the sachets produced in Niamey and follows the commodity and value chains we describe in the following sections. In addition, three companies produce water sachets on an industrial scale: Souley Group S.A., Mataba Torodi International, and Niger Lait S.A. We will revisit them in Chapter 6. Like the urban water regime in Jakarta, sachet water is just one element in a complex "scattered archipelago" of water access in Niamey (Kooy and Bakker 2014: 63).

When the piped system failed to provide residents with access to water, hybrid systems emerged in which the division between private and public (and community) entities is blurred. The sachet water economy is integrated into this inequitable system of water access. Compared to *ga'ruwa*, who have existed in Niamey for decades (Youngstedt et al. 2016), sachet water sales are relatively new. Initially in the 1990s, sachet water producers and sellers manually filled small bags with water and hand tied them. Today, however, much of the sachet water in Niamey is produced through a mechanized process: "the latest low-cost technological incarnation of vended water" (Stoler et al. 2012: 3).

Because producing and vending sachet water spans government, private, and community sectors, it is difficult to pinpoint exactly when it became part of Niamey's cultural landscape. In the absence of official records, we rely on our interviews with producers, vendors, and customers. The practice of manually filling plastic bags (still commonly used for selling ice) with water and hand tying them is older than the current mechanized process. In their study of sachet water vending in Accra, Ghana, Stoler et al. (2012) note the emergence of automated heat-sealing sachet machines in the late 1990s, all of which were imported from China.

In Niamey, the automated system appeared more recently than in Ghana because the machines used in Niger were first imported from China to Nigeria, from where the concept and the technology spread to Niger. (Nigeria continues to import water-bagging machines from Chinese companies but now also manufactures its own machines. Chinese companies include Dingli Packing Machinery and Zhangjagang Sanofi Machinery Co., Limited, while Nigerian companies include Global Sterling Products Limited and Business Cell Nigerian Limited. Sachet water machines are not manufactured in Niger.) This diffusion explains why the most commonly used term in Niger for sachet water is the English phrase "pure water," despite the fact that French is the official language in Niger. The term is pronounced as *piya wata*—a Hausification of English.

The most experienced water vendors that we interviewed indicated that they had been selling modern sachet water for between ten and thirteen years, although we must consider that the sachets themselves were initially imported from Nigeria into Niger before the automated technology arrived. In other words, we can say with confidence that modern, machine-sealed sachet water has been available in Niamey since the early 2000s, but it is difficult to determine exactly at what point residents of Niamey began machine-sealing the bags. Furthermore, machine-sealed *piya wata* did not immediately supplant the hand-tied bags. For six to eight years they shared the market. Today, hand-tied bags of water are rare in Niamey. "Pure water" sachets are now an integral part of a larger, complex landscape of water access and global trade in the city.

The "Pure Water" Backstory: A Commodity Chain and Value Chain Analysis

We now construct the backstory of plastic water bags by moving from the discarded bags; to those who consumed the water in them and discarded them; through those who produce, refrigerate, and sell them; and finally to the larger system of trade to which sachet water is connected. Because the production of sachet water is typically not visible from the street, we consulted "gatekeepers" who knew where the water production was taking place. We knew that most of the activities in the sachet water economy occur in low-income neighborhoods, where water access is more tenuous, a condition also noted in Stoler's (2012) study: thirteen of the seventeen neighborhoods included in this study are poor. We did, however, consider how the sale of sachet water plays out in wealthy neighborhoods, and thus we included water vendors in three of these as well.

Plastic Water Bags

We observed discarded plastic water bags in most neighborhoods in Niamey, wealthy and poor alike, indicating that the consumption of sachet water and the habit of discarding the bag immediately after consumption spans the socioeconomic spectrum. The density of discarded bags, however, was greatest in poor neighborhoods, because, unlike in wealthy ones such as Issa Beri, residents there cannot afford to pay private workers to regularly collect trash from roadsides and compounds. Because sachet water is the only form of packaged water sold cold and on the street, it is usually consumed immediately upon purchase. Unlike plastic bottles, which are reused, there is little use for the empty sachet without an advanced recycling facility that is capable of transforming the bags into new commodities. In the face of inadequate solid waste collection, most people discard the bags wherever they happen to be after drinking their contents.

The discarded bags ultimately clog gutters, sewers, and drains, increasing exposure to raw sewage, animal waste, and other toxins—a situation mirrored in Stoler et al.'s (2012) study of Accra. Those that are collected from wealthy neighborhoods are deposited in empty areas or poor ones on the urban periphery. (Plastic grocery bags are collected from wealthy neighborhoods and deposited in much the same way.) In addition, plastic bags are burned in trash heaps, causing air pollution. The environmental impact of discarded plastic bags and relocated trash was particularly evident during our fieldwork, which twice overlapped with the rainy season in southern Niger. Furthermore, since the storm sewers of Niamey empty into the Niger River, many bags end up in the city's water source, threatening fish and potentially clogging pipes that draw water into the city's water treatment facilities.

The clear plastic bags are printed with text indicating the reputed location of their production and other messages. We collected sixty-five different brands of sachet water and noted many others. More were produced in Nigeria than in Niger, according to their labels. Most bags made in Niger include some English, including one that is entirely in English: "Do Freedom Water." Nigérien-made products rarely include English on their labels. The "Do Freedom Water" sachet indicates that it was produced by "Shehu Foods, Banizumbu, Niamy [sic]," with the spelling of "Shehu" and "Banizumbu" reflecting Anglophone Nigerian conventions. Francophone Nigériens spell these names as "Cheifou" and "Banizoumbou." This suggests that Nigerians are working in the water-producing business in Niamey, or at least that some Nigerians are producing the ink lettering that is stamped on plastic water bags. The "D.A.I.B.A." brand bag describes its contents in French, Hausa, and English: "Eau Filtree," "Tatatchen Ruwan Sha," and "Natural Water," respectively.

Most brands made some claims about the quality of their product. The "Nouveau Eghazar-Water, Eau de Table" brand manufactured by Ets. Abdel in Niamey claims that its water is "Traitement Fait Par Un System," "Ultra Norderne [sic]," "De Filtration Avancee [sic]," "Mise en Sachet," and "Entierement Securise [sic]" ("Machine Made,"

"Ultra-modern," "Advanced Filtration," "Put in the Bag," and "Entirely Secure"). Most brands include the words "Manufacturing Date" and "Best Before," but none had stamped dates. Most brands also include the words "Dispose of Properly" and small images of a person dropping the bag in a rubbish bin. Adding a brand stamp to clear plastic bags adds value, as this process distinguishes contemporary sachets from the earlier, hand-tied ones. However, it is merely the presence of the label itself that adds value and communicates ideas of the water's "purity," not the actual text of the label. In our research, there appeared to be no brand loyalty among customers. We return to the topic of branding when we discuss the role of consumers in the sachet water economy and again in Chapter 6.

Consumers

People who purchase plastic bags of water do so to fulfill the desire to consume water that is cold and perceived to be clean, as the sachets are sold from coolers on pushcarts or directly from refrigerators in boutiques or small shops. In other words, it is the cold temperature, the perception of purity created by clear water and a machine seal, and the label (all elements of production) that add value to sachet water for consumers. Consumers in Niamey are accustomed to purchasing all kinds of counterfeit or pirated products, including clothing, wristwatches, mobile phones, shoes, CDs, and cassette tapes. As Stoller (2002: 95) points out, "In the mass consumer market, difference in quality between the original and the copy is sometimes negligible." Sachet water falls into the category of "cheaply manufactured goods whose only allure is the fame of their trademark" (Stoller 2002: 95).

The trademark or label becomes more important than the quality of the item (Coombe 1996; Stoller 2002). However, in the case of *piya wata* in Niamey, any sachet label will do. As far as we can tell, consumers do not have brand loyalties. No one—not producers, not vendors, not consumers—mentioned loyalty to or preference for any particular brand. We never saw a consumer read the brand logo or lettering. All that seems to matter is that there is a brand stamp on the sachet and that it is cold.

The desire for cold water seems natural in one of the world's hottest cities, but we maintain that this desire has only recently been created in Niger through the marketing and commoditization of water. In traditional Hausa thought, it was seen as necessary to keep cold and hot in balance to ensure health, but "the effects of heat (*zafi*) are felt mainly on the external body and are generally not as important in pathological explanations as are the effects of cold (*sanyi*)" (Wall 1988: 190). Cold produces dead blood, increased phlegm, colds, stomach ailments, rashes, arthritis, headaches, lethargy, and infertility (Wall 1988: 187–90). The Songhay and the Zarma similarly recognize hot and cold illnesses (Nowak, personal communication; Olivier de Sardan 1982; Stoller 2016 [personal correspondence]; Stoller and Olkes 1987). Cold illnesses—called *yeni* in Songhay—are associated with rheumatism, arthritis, upper respiratory infections, and other similar disorders (Stoller 2016 [personal correspondence]). Furthermore, there is a spirit dimension to coldness: cold spirits—called *hargay* in Songhay—are spirits of death and disorder (Stoller 2016 [personal correspondence]). The Zarma, who, like the Hausa, have adopted the Songhay cold spirits (*hargay*), associate cold with the same ailments that Hausa do, except that they do not regard it as a cause of stomach ailments and infertility (Nowak, personal communication).

In short, people living in Niamey hail from backgrounds wherein cold, including cold water, was viewed negatively. Were such feelings to persist or be pervasive in the present day, the marketability of sachet water would be restricted. The young, however, associate access to refrigerated consumer products with modernity—and the consumption of cold sachet water is largely restricted to those who are below forty years of age. Furthermore, many of them are less intolerant of cold drinks, having suffered less tooth decay than older people, many of whom have dental problems due to the popularity of highly sugared tea and coffee and the lack of effective oral hygiene practices.

The process of commodifying the commons typically leads to individualism, and this is the case with sachet water. Traditionally, Nigériens consumed water quickly and all at once after meals, usually from shared receptacles that have evolved over time—gourds, clay cups, stainless steel cups, and plastic cups. In contrast, Nigériens drink *piya wata* in between meals, especially in the hot season, from individual sachets, though these are sometimes shared too.

The consumption of sachet water is typically associated with poverty and inadequate access to water (Stoler et al. 2012), but our research indicates that both poor and rich use the product with different associations. A one-half-liter sachet of cold water sells for 25 FCFA (approximately US$0.05 in 2015) and is hence relatively affordable, but it still costs much more than the same amount taken directly from public standpipes, which is the cheapest way for Niamey residents without a direct connection to the piped network to obtain treated water.

The people we interviewed in poor neighborhoods saw the purchase and consumption of sachet water as a visible display of their poverty because this typically indicates that they lack access to refrigeration. As one participant explained, "The only time I can drink cold water is when I buy *piya wata* because my family is too poor to own a refrigerator." Another participant told us, "When I drink *piya wata*, everyone knows I am poor." (People also purchase "pure water" due to simply being far from home due to work, socializing, or shopping.)

Sachet water was also consumed, with rather different connotations, in three wealthy neighborhoods we observed (Dar es Salam, Issa Beri, and Maurice Delens), where most households are connected to the electrical grid and the piped water network and have backup generators. First, we noticed that sachet water is sold from boutiques in these neighborhoods, implying a customer base for this product, hence our inclusion of them in this study. In one instance, attendees at a wedding (where large tents were erected on the street for the day) made multiple trips throughout the day to a nearby boutique to purchase sachet water. In another case, the host of a party bought sachet water in bulk (bags of twenty sachets sold at air temperature), kept it in a refrigerator at his home, and then served it to guests during the party. Finally, a man driving a relatively new Mercedes-Benz car pulled up to a boutique, bought a sachet of water, consumed it immediately, and then discarded the bag on the ground outside the boutique as he drove away.

In the consumption of sachet water by the wealthy, various symbolic and practical aspects are evident. A need to quench thirst and a desire for cold water are conditions shared by the wealthy and poor alike, but while poor consumers we interviewed felt that the act of consuming sachet water was a display of poverty, wealthier consumers tended to see sachet water and its accessibility in public areas as a convenience to which they are entitled. Furthermore, the discarding of bags by wealthy consumers in wealthy neighborhoods can communicate their status as people who can afford trash collection, in contrast to poor consumers, who simply have no other place to put the empty bags. Consumers pay the same price for sachet water across the city. The impact of that cost on the individual, however, is different depending on the socioeconomic position of the consumer. Furthermore, the symbolic value of the sachet varies by consumer group.

Vendors

The consumers, who ultimately discard the plastic water bags and create the landscapes of waste, purchase those sachets in one of two

Figure 5.1 Mobile sachet water vendors, with a cooler containing bags on a pushcart in the foreground. The girl on the left has sachets in the bucket on her head. Photo by the authors.

primary ways: from mobile vendors who walk the streets selling the cold bags out of coolers on pushcarts or from buckets carried atop their heads (Figure 5.1) or from small roadside boutiques that sell the bags out of refrigerators (Figures 5.2 and 5.3). Interviews with both types of vendors further reveal the backstory of discarded plastic bags and indicate considerable complexity in the economy of water vending. Vendors are an important link in the commodity chain between consumers and producers.

The most publicly visible vendors of sachet water in Niamey are the (constantly) mobile ones and those with roadside tables. The latter are also mobile in the sense that they store their tables in their homes during the evening and wheel them to their regular roadside spots during the day. Most of the mobile vendors work in poor neighborhoods, and most are children or young people—as mobile water vending requires agility and physical labor, including hours of walking in typically hot temperatures—whereas the sedentary ones operating at tables are adults. Selling *piya wata* on foot is an entry-level job for local and recent migrant youth in Niamey. Mobile vendors in our study ranged from eight to forty years old. Most are boys and men.

Figure 5.2 A boutique with sachet water in bulk on the patio. Photo by the authors.

Fifty percent identified themselves as Hausa, and 50 percent identified themselves as Zarma. They had been in business for as little as three days and for as long as nine years, with an average of fourteen months. Five youths explained that they work long hours only during school vacations. Both types of vendors work long days: eight hours on average, six or seven days per week. When we asked mobile vendors why they sold water, three common answers emerged: they were instructed to do so by family members, they were unable to find other work, or they preferred it to sitting around "doing nothing" (*zaman banza* in Hausa).

The mobile vendors in our sample almost always worked for people who owned refrigerators. In contrast to the economy of *ga'ruwa* water delivery, which is entirely controlled by men, women in Niamey, like Mariama in the opening vignette of this chapter, have creatively—and virtually invisibly—inserted themselves into the commodity chain by purchasing sachets in bulk and refrigerators and then sending their children out to sell the cold bags. Adult women, especially those who are married with children and extended families, cannot spend eight or more hours a day walking the streets to sell water.

Figure 5.3 A *boutiquier* and his refrigerator filled with cold sachet water for sale. Photo by the authors.

However, women who have acquired refrigerators, even small ones, can buy sachet water in bulk and send their children out to sell it for 25 FCFA per bag, a standard price throughout the city, yielding a gross profit of 350 FCFA per bag of twenty. (Shop owners pay producers more than women with refrigerators at home—200 FCFA on average for a bag of twenty sachets—for reasons that we could not determine). A few vendors sell warm sachet water for 15 FCFA each. Although many of the mobile vendors we interviewed could not estimate the number of sachets they sold per day on average, those that could estimated that they sell as few as forty bags a day and as many as 350 bags per day, with an average of 122 water sachets sold daily. Thus, a woman whose child sells 122 bags of *piya wata* in a day earns 2,135 FCFA or US$4.27 in gross profits. As in Mariama's case, women's income is offset by the costs of purchasing refrigerators (secondhand refrigerators are available for 25,000 FCFA in Niamey), paying higher electric bills, and paying children who are not their own to sell sachet water.

Although women are newcomers to the water economy of Niamey, their entrepreneurial strategy draws from older patterns. Many scholars (Callaway 1984; Hill 1969; Schildkrout 1982; Wall 1988) have documented a long history of women in the region—even secluded wives—operating successful businesses from their homes by sending their children into the streets and markets to sell their products, particularly cooked food and kola nuts. "In Hausa society, any income that wives earn on their own is inalienably theirs" (Callaway 1984: 440), and women in Niamey need money more than ever to survive in the capitalist economy, to participate in the naming ceremonies and weddings of relatives and friends, and to integrate themselves into women's exchange networks (*foyandi*). In other words, women sell water indirectly and take advantage of the demand for water to support their families, an activity that, if done directly, would not be culturally acceptable.

Of the twenty mobile sellers we interviewed who worked for family members, only three were paid directly for their work: an average of 11,000 FCFA per month. The others were paid in kind, with money from the sales used to buy them food or supplies and clothes for school or important Muslim holidays or for transport. Those who worked for nonfamily members were paid an average of 8,000 FCFA per month for their work.

There was consensus about the difficulty of the work. Several cited fatigue and "heavy" legs due to pushing heavy carts on the uneven, sandy streets of Niamey. Those who carry the sachets in bowls atop their heads explained that they suffer frequent headaches. Nevertheless, most mobile vendors maintained that they were proud to have been given the responsibility to work and handle money, to help support their families, and to contribute to their own needs.

The second group of sachet water vendors—*boutiquiers*—sell sachets from shops on the roadside and thus are sedentary rather than mobile.

(The most common goods for sale in the boutiques of Niamey are canned tomatoes and fish, rice, cooking oil, powdered milk, tea, coffee, cigarettes, mobile phone cards, and sachet water.) All twenty-seven boutique owners in our study were men, with an average age of thirty-six years. Most were Hausa, but Zarma, Songhay, Fulani, and Arabs are also involved. *Boutiquiers* choose the products to sell in their shops, unlike children who sell sachet water on pushcarts because they are told to do so by their mothers. They say they sell sachet water because it is popular and many people buy it: a response that was universal across all *boutiquiers* in both wealthy and poor neighborhoods.

Boutiquiers purchase sachet water in bulk (bags of twenty sachets for an average of 200 FCFA) from producers who deliver it regularly to the shops. Some then sell the bulk bags to the mobile vendors with carts. Most boutiques in wealthy neighborhoods sell bottled water alongside the sachets, but *boutiquiers* noted that the latter is much more popular, enjoys higher consumer confidence, and sells faster. In poor neighborhoods, *boutiquiers* purposefully carried no bottled water, stating that it was too expensive for their clientele; one stated, "Rich clients buy bottled mineral water. Most of my clients buy *piya wata*." Another explained, "Everyone likes it. It is cheap. My clients are mostly poor people."

As mentioned earlier, there was no brand loyalty among *boutiquiers*, and they did not mention any preference for labels. Rather, the loyalty lies between the *boutiquiers* and the sachet water producer. All *boutiquiers* told us that they buy sachet water in bulk from the same one or two producers, who regularly deliver the water to the shops.

Although neither mobile vendors nor *boutiquiers* were inclined to estimate the number of sachets they sell per day, some reported selling as few as 70 and as many as 400 sachets per day. Furthermore, responses revealed a significant difference between wealthy neighborhoods, where *boutiquiers* estimated that they sold about 150 sachets per day, and poor ones, where about 350 were sold. Sales volumes occur in an inverse relationship to refrigerator ownership. Among the twenty-seven *boutiquiers* in the study, ownership ranged from one to five refrigerators, with an average ownership of two refrigerators each. *Boutiquiers* in wealthy neighborhoods owned twice as many refrigerators on average as those operating in poor neighborhoods—2.8 compared to 1.4.

This disparity in sales can be explained at least in part by the fact that *boutiquiers* in poor neighborhoods devote a much higher percentage of their refrigerator space to *piya wata* than *boutiquiers* in wealthy neighborhoods do. Whatever the neighborhood, many more bags are reportedly sold per day during the hot season (three to four months) than at other times of the year. Sachets sold in boutiques are usually consumed immediately upon purchase and then discarded nearby, usually in front of or around the boutique itself. In fact, one of the ways we identified boutiques that sold sachets of water was by observing the density of the discarded bags outside the shops.

Almost three-quarters of *boutiquiers* explained that their work is difficult, but their problems seem relatively minor compared with those of the mobile vendors. Broken sachets and constantly getting up and down from their chairs to serve customers were the most commonly cited problems. Others mentioned slow sales during the cold season (three to four months), long work hours, and the constant need to sweep up discarded sachets.

Producers

While cold water vending is common in both wealthy and poor neighborhoods in Niamey, sachet water producers are almost exclusively located in poor neighborhoods. Sachet water vendors in Niamey buy the sachets in bulk (as described above) from producers. For interview purposes, we divided these producers into two groups: small-scale and large-scale producers. (Three industrial-scale sachet water producers also operate in Niamey, as noted earlier.) We interviewed only one in the former category—a sixty-year-old semiretired Hausa man living in Zongo—but he shared information about the experiences of other small-scale producers that he knows. Small-scale producers buy rolls of plastic in the market. Then, using simple electric machines about the size of laptop computers that are imported from China and available in local markets for about 5,000 FCFA, they create bags, fill them with water stored in their homes, and then manually seal them, often suffering repeated small but painful electric shocks from the crude machines. Since they produce only about one bag per minute, they cannot compete with large-scale producers. This method is even slower than the hand-tied method in which no machines are employed.

The small-scale producer, in a practice he told us was common, buys prelabelled *piya wata* sachets and fills them manually with water delivered to his home by *ga'ruwa* and stored in a clay pot. These sachets are indistinguishable from those that take water directly from the piped network, filter it, and pack it using automated machines. However, for reasons discussed earlier and to be further elaborated below, only a very small number of consumers purchase the cheaper, unlabeled, hand-tied bags of water that sell for 15 FCFA each. Most are willing to pay 25 FCFA each for sachets of labeled *piya wata* as they trust that this water is "proper" and "pure."

We interviewed nine large-scale producers, an even mix of Hausa and Zarma, who had been in the business for between three months and six years. These producers have purchased machines that are connected to Niamey's piped water system. These machines are about two meters tall, one meter wide, and one meter deep (Figure 5.4).

Treated water comes through the piped network to the homes or shops of the producers and passes through three filters before it enters the machine, which automatically fills one-half-liter plastic bags with water and seals them (Figure 5.5). Producers can make approximately thirty sachets per minute with this automated method. Individual sachets are then manually put in larger bags of twenty and sold to boutiques in Niamey or delivered to small shops in surrounding villages. A few sell sachets in bulk to the vendors with pushcarts or to women with refrigerators who send their children out to sell water.

Among the nine large-scale producers we interviewed, five produced sachets in small storage spaces or roadside shops, and four had installed the machines in their homes. Since work done in the home is not taxed, the latter serves mostly as a means to avoid paying government taxes. Large-scale sachet production is often a family affair, partially as a result of the space of production. Children help run the machines or pack the sachets in larger bags of twenty. In other cases, producers employ and pay others who may or may not be family members. Some owned trucks that they used to deliver sachets in bulk to boutiques and villages. Income from large-scale production varies depending on the number of clients.

Estimating profits for producers is complicated. Producers who sell an average of 1,400 one-half-liter bags per day—in batches of twenty for 200 FCFA per batch—earn approximately 14,000 FCFA per day. However, producers have more expenses than vendors. First, large-scale producers had to buy the machines, mostly buying them in Nigeria since they were not yet available in Niger, at a cost of around 1.33 million FCFA. This did not include the expense involved in travel to and from Nigeria, transporting the machine back to Niamey, and customs fees. This is a considerable amount, especially if one considers that the average Nigérien earns approximately 500–1000 FCFA per day and the producers are not wealthy.

One producer stated that it took him three years to pay off the loan to buy his machine. Furthermore, producers complained that

Figure 5.4 Automated sachet-filling machine in a private home. Photo by the authors.

the machines were always breaking down, particularly the ones imported to Nigeria from China, forcing them to pay for repairs and replacement parts. In addition to these up-front costs, the machines are connected to the piped water system, which means paying for the connection and for monthly water use. Electricity is required to run the

"Pure Water" in Niamey • 115

Figure 5.5 Close-up image of the automated process. Photo by the authors.

machines, and a few producers indicated that they had also purchased generators so that they could continue to produce sachets even when the flow of electricity was inconsistent. Generators allow producers to continue production during the hot season, which is when producers reported their largest profits but when the electrical supply to Niamey is least reliable. Producers also have to buy rolls of plastic

Figure 5.6 Sachet water in bulk (bags of twenty) stored in a garage waiting to be delivered to boutiques in Niamey and the surrounding villages. Photo by the authors.

bags from the market to use in the machines and pay any employees they have (Figure 5.6).

Producers expressed varying levels of satisfaction, with all citing some difficulties in their occupation. Two told us that the income was acceptable; were it not so, they would seek other work. Three said that they were actively looking for other jobs because there is little money left over from water sales after paying bills. Other problems included working with unreliable machines, painful hands due to handling cold water for hours at a time, stiff competition, and slow sales volume outside the hot season.

Sachet water producers are key players in the hybridized water economy in Niamey. The machines they use are connected to the piped water network and electrical grid, but the government does not collect taxes on the sachets sold or profits made. These producers also link local vendors to global systems, as the machines they use are made in Nigeria or imported from China to Nigeria, and most (if not all) of the plastic bags are made outside Niger as well. Thus, the landscape backstory of discarded plastic water bags does not end with producers;

rather, it continues beyond them and across a multitude of systems of global trade and interdependence.

Purity: Bottles or Sachets

But how pure is "pure water"? Our research on sachets bears out the findings of Gleick (2010) and others, who question the idea that bottled water—which is less regulated than tap water by national governments—is cleaner and safer than tap water. In theory, the purification and filtration should guarantee that the water is pure. In addition, the mechanical sealing, which indicates industrial production, adds value to the sachet and increases consumer confidence in a way that the hand-tied bags do not, but the sachets remain more affordable than their bottled equivalent. In practice, matters are more complex. Producers are not required to have their filters tested. None of those we observed in the cottage industry wore gloves, masks, or uniforms. We encountered a young man who was working with a bad finger cut, an injury he sustained while using the sachet-sealing machine.

Furthermore, consumers have no way of knowing the sources of the water in the sachets. They assume that it comes directly from the piped network. As long as the sachets have printed labels and the water appears clear, consumers typically assume that the water is pure. We could not identify any studies of the purity of "pure water" in Niger, but tests of "pure water" in neighboring Nigeria are alarming because "the results of most of the studies on sachet water to determine purity and safety have almost always churned up evidence of microbial and in some cases chemical contaminants" (Nwadike 2012: 2; see also Dada 2009; Ngmekpele and Ephraim 2014; Stoler et al. 2014; Fisher et al. 2015; Ibrahim et al. 2015). One *boutiquier* in our study suggested with a mixture of cynicism and resignation that there was no point questioning him about the purity of his water. "This is Africa," he said. He knows that he cannot guarantee the purity of his *piya wata*, and he knows of clients who have become ill from consuming it, but he must continue selling it to earn a living. He added that his clients' trust in the purity of sachet water is naive.

In addition, the plastic bags themselves pose health risks. Although consumers of sachet water typically drink the water immediately, producers make and store the bags in garages or homes for days or weeks before they are distributed. *Boutiquiers* who buy water in bulk may store it for weeks or months in their shops until it sells. Garages, homes, and shops all usually lack air conditioning and are subject to sweltering heat during most of the year. The length and conditions of sachet storage can cause the plastic to leach into the water it holds (Hungerford 2012). These concerns do not seem to impede the water sachet business in Niamey, however, which is instead affected by other conditions, such as flow of water, electricity, temperatures (producers and vendors reported reduced profits during the cool season), or Muslims abstaining from food and water between sunrise and sunset during the month of Ramadan.

Sachet water and the discarded bags are part of a larger system we refer to as "global plastic capitalism" (Keough and Youngstedt 2014). This system involves the production and international trade of plastic materials—including water bottles, bags, shoes, kitchenware, and toys, among many other items—as well as the perceived and planned obsolescence that maintains and expands the customer base for particular products. Sachet water, like bottled water (Gleick 2010; Hawkins et al. 2015), is linked to the global plastic capitalist system. Like plastic bottles, sachets are made from cheap plastic produced in mass quantities and transformed into products that, when filled with water, suggest to consumers that they are drinking clean water. Like bottled water, sachet water emerged as water systems became privatized through neoliberal economic practices, fragmenting the water supply and affecting the waste infrastructure (or the environment, when such infrastructure is absent, as with Niamey's landscapes of discarded water bags).

Bottled and sachet water have some characteristics in common, but they are also different in several ways. The consumption of bottled water is a status symbol, largely because it is very expensive. Most Niameyans have never tasted it. The consumption of sachet water is a symbol of poverty to some but comes with ideas of entitlement to others. Consumers consider both bottled water and sachet water forms of pure water, in part because both forms are mechanically sealed, but, in reality, the former is probably cleaner than the latter. However, sachet water can be quite clean if the production process is done properly. The biggest difference between the two forms of packaged water, though, is that, in Niamey, the plastic bottles—unlike the sachets—are recycled informally (women in particular use them to store cooking oil and other cooking ingredients). While much has been written on the environmental hazards created by discarded plastic water bottles, it is the sachets that are a bigger waste and drain-clogging problem in Niamey.

Conclusion: Commoditization and Value

This chapter has examined the backstory of sachet water in Niamey — a complex and fluid example of global plastic capitalism in motion. By linking global value chains and commodity chains, this story reveals important dimensions of power, class, gender, age, cultural values, and concepts of purity in Niamey. Our examination of global value chains and commodity chains involved tracing discarded sachet water bags on the roadsides of Niamey back to the consumers who dropped them, the vendors who sold them, the producers — using the municipal water supply and electrical grid — who filled them, and the Chinese and Nigerian companies that manufacture and export plastic bags and automated sachet water–packing machines. We demonstrated how the importance of labels, temperature of water, time of year, and ideas about water purity add both economic and symbolic value to the sachet, typically during the stages of production. Finally, we explained how the sachet water economy simultaneously combines characteristics of several water governance models.

The commoditization of water and the backstory of discarded water bags is not complete, however, without careful consideration of the commodity chains and value chains of which it is a part. These chains connect two distinct commodities: water and plastic bags. Each travels through a different commodity chain until the point at which they are mechanically combined. When the plastic bag is filled with (presumably) filtered water and sealed in a labeled bag, a new, more expensive commodity is created, one whose economic value increases further upon refrigeration, especially during the hot season, and whose symbolic value varies with the socioeconomic status of the consumer.

It is here that we return to our comparison between sachet water and cash crop value chains that we mentioned earlier. While the value of cash crops is added and subtracted at different stages in the commodity chain, the packaging does little to change the value of the commodity, other than perhaps to protect it in shipping. Sachets, however, markedly increase the value of the water contained within them. Unlike cash crops, which are grown primarily for export out of the continent, sachet water is consumed very close to the place where it is produced — sometimes on the same street — largely due to the fact that water is heavy and expensive to transport. People at all socioeconomic levels in Niamey, including the producers and their workers, consume sachet water; therefore, there is little "alienation" along a relatively short commodity chain. Sachet water is linked to the global economy through the one-way flow of manufactured products from China to Africa (Lee 2014), but, unlike cash crops grown in Africa and bound for Europe or North America, the consumer base for sachet water is very much a local one.

The points at which value is added or subtracted from sachets result in further differences from cash crops. Unlike cash crops, whose value increases continuously from raw material to consumer, the value of sachets and the water in them fluctuates as they move through the commodity chain and with the time of year. This is in part due to the fact that, unlike agricultural products, water can be stored in sachets for long periods of time, minimizing the risk of investment to producers and vendors and guaranteeing a consistent supply to consumers.

How might the theories of Tsing, Appadurai, and Swyngedouw further illuminate questions of value in the selling of sachet water? Our study supports Tsing's assertion (2013) that value is created and assessed at all points in the commodity chain, and that noncapitalist social relations, such as those between refrigerator-owning mothers and their children, change in part due to their involvement in capitalist processes. Appadurai (1986) emphasizes that the process of commodity exchange is key in creating value, and in this case sachet water is exchanged twice in the commodity chain: once between producer and vendor and then again between vendor and customer.

Although the vendors pay for the filled bags, the bag's economic value increases once it is cooled, so processes occurring before the second exchange add value as well. It is the process of production, however, as Swyngedouw suggests (2004), that results in the most significant change in economic value for sachet water. When the two individual commodities are combined to produce *piya wata*, the economic value of each increases significantly, not only as a product ready for consumption but also as product that can be stored long term. Furthermore, production itself offers a livelihood for men in a city with high unemployment. The process of chilling water is another phase of production that increases the economic value of the new commodity exponentially. In fact, cold water is so valuable to consumers that they will choose it despite perceptions about its public consumption and cultural associations with cold-related health problems.

Producers, vendors, and consumers benefit from the fact that sachets are used to store water. For producers, this allows them to produce a surplus of sachets when water flow through the piped network is consistent, ensuring a continued supply of sachets even when—as during the hot season, when demand for sachet water is also highest—the piped network later proves unreliable. Vendors reported selling more sachets during this season than at any other time of the year, and many vendors purchased extra bulk packs of sachets in anticipation of this demand. Of course, the fluctuating electrical supply during the hot season presents challenges for refrigeration. To consumers, the availability of cold water is even more valuable during the hot season, when daytime temperatures reach 45°C or higher and both formal and hybrid systems of water delivery are inhibited by reduced water flow.

While water sachets are part of larger, formal trade flows, they are also integrated into Niamey's "community economy" (Bakker 2010)—the sector that employs most of the city's population (Youngstedt 2013). In Niamey, the sachet water economy has remained largely unregulated by the state or private entities. There are indirect controls, such as the price of water obtained through the piped network and the charge for electricity, that are set by the government or public-private partnerships, but the distribution and sale of sachets remain under the control of producers and vendors, and these actors standardize the price (both for individual sale and in bulk) of sachets throughout the city.

The sale of sachet water moves between government, corporate, and community economic systems, a condition accounted for in Bakker's (2010) typology of the urban water supply. Bakker locates water vendors in vague water service areas between those under corporate control and those entirely outside both corporate and government control. Large-scale producers of water sachets are connected to water and electrical distribution systems, both of which are controlled by public-private partnerships, but many producers do not pay taxes on the sachets sold because their production occurs in their homes. Furthermore, the producers in this story have privatized part of Niamey's water industry, both in their role as individual entrepreneurs and in the sense that many produce sachet water in the privacy of their home. However, when sachets are sold from roadside boutiques, it is *boutiquiers* (rather than producers) who must pay taxes on the sachets and on all other goods they sell from their stores. The mobile vendors and their mothers and grandmothers, on the other hand, pay no taxes on the income they collect.

Furthermore, producers and vendors alike benefit from informal social networks of regular customers, which, Tsing (2013) asserts, exist when alienation between worker and commodity is minimized. The social networks among producers, vendors, and consumers are, according to our informants, vital for business, just as these social networks advantage those in other informal economies in Africa (Meagher 2010). Thus, the sachet water economy is a quintessential example of Bakker's (2010) "hybrid" economy because it combines elements of all three governance models: public-government, private corporations, and community organizations. It exists because of direct access to the piped water and electrical networks, which are functions of public-private partnerships, it serves community needs (the desire for both cold water and employment), and it involves social networks for efficient function, but there is no cooperative community governance of the system, despite the fact that the price of a cold sachet is fixed throughout the city.

In sum, the political economy of "pure water" produces an extensive commodity and value chain through which the bags pass. These exchanges are embedded in cultural contexts, as there are commonly held norms about who can sell water in public spaces and changing concerns about consuming cold products. The age-based and gendered division of labor within the process, the cultural perceptions placed on the consumption of sachet water, the social networks involved in the production and distribution of "pure water," the hybridized state in which the "pure water" system is positioned, and the global conditions that affect local realities are all part of the backstory of discarded plastic water bags. These are essential elements in the biographies of objects discarded on the roadsides of Niamey.

Note

This chapter was first published in an earlier version in *Africa: Journal of the International Africa Institute* 88(1): 38–62, 2018, under the title "'Pure Water' in Niamey, Niger: The Backstory of Sachet Water in a Landscape of Waste." Republished with permission.

6

Fluid Materiality in Niamey

In September 2016, when we arrived in Niamey for the beginning of a year of fieldwork, one of our first tasks was to find housing. Having spent years (and, in Youngstedt's case, decades) doing research in Niamey, we were already familiar with the city, its neighborhoods, and most of the housing options within our budget. While we desired some modern conveniences, like indoor plumbing, a direct connection to the piped water network, and electricity in our home, as water researchers, we wanted to avoid what we considered to be excessive water-based luxuries in the Sahel, like a grass lawn or a swimming pool. So, although we had been studying water access and material culture in Niger already, it was really the process of securing housing that made us think practically, and more seriously, about the materiality of water.

The house we ended up renting, in the ORTN neighborhood of Commune II in Niamey, was a colonial-style house embellished with traditional Hausa iconography in a mixed income neighborhood not closely affiliated with the larger expatriate communities in Plateau I and Kouara Kano. Our house was across the street from the South African Consulate and two doors down from an opulent villa occupied by Chinese businessmen. In contrast, squatters living in makeshift grass houses covered with plastic sheets occupied the lots next to and behind us. The black plastic bags that solidify commercial exchanges in Niamey (even when purchasing a single item), which have become ubiquitous in the Nigérien landscape over the last twenty years, and discarded used water sachets, collected in unoccupied spaces at the ends of our street. Our occupancy in this space, one that included formally established plots of land but whose occupants were not always the owners and did not always have permanent structures on the property, fell between formal and hybrid forms of the reality of utility access in Niamey.

In our home, we had direct connections to the city's electrical and water-sewer networks and enjoyed the conveniences of indoor plumbing, but we allowed those living (we presumed) informally behind us to splice our electrical connection to run their television in the evening, and we shared water with the neighbors next to us, whose children would come to our compound every few days with the ubiquitous 25-liter plastic jugs to fill from the hose and water tap in our yard. In other words, we participated in the informalities of living in Niamey to which so much of the population is subject by expanding the territories of access for our neighbors. In this way, we created new versions of the hydrosocial territories described in Chapter 1.

Furthermore, as residents and home renters in Niamey, we also became customers of the water utility company and were responsible for paying monthly bills determined through the process of block pricing. While our housing allowance provided by the Fulbright Foundation covered basic utilities, we became more conscious of the role we played in Niamey's water economies not just as researchers but now also as residents.

Finally, we became acutely aware of our material involvement in the water economy. As a country whose economy falls largely in the primary sector (one based on natural resource extraction and agricultural exports), Niger has little domestic industrial production, and most consumer goods have been imported from China and, to a lesser extent, European countries. The house we were renting was occupied by another scholar we knew, and prior to our arrival, he had installed an imported water filtration system in the kitchen so that we could filter piped water to drink and for cooking. In addition to the kitchen, three sinks, three toilets, and two showers provided other water network connections, and we had an outdoor water tap in our yard. We kept tap water in recycled plastic bottles on hand in case of water cuts, which turned out to be less common in our neighborhood than in others, and as Americans poorly adapted to dry conditions, we carried reusable water bottles with us whenever we left the house.

Several houses in our neighborhood had external water storage tanks holding a few hundred gallons of water each, usually elevated to take advantage of gravity, in case of reduced water provision. In addition, we occasionally purchased *piya wata* when out in the city either for ourselves if we ran out, or for friends we were with, especially during the hot season. As much as we tried to avoid it, cheap Chinese plastic imports facilitated our participation, and we realized how easily these plastic materials had come to replace domestically produced goods like the clay pots made in Boubon or the leather sandals made throughout the country. In fact, when traveling to Boubon several times over the last ten years, we noticed fewer pots being made, more pots stockpiled in unoccupied spaces, and fewer pots being sold in the Niamey markets that used to specialize in these types of products. In short, global plastic capitalism was alive and thriving in Niamey.

We touched on the materiality of water in the previous chapter on sachets. This chapter expands on the material elements of water, including its packaging and branding, by considering how water's materiality manifests itself in Niamey's visible landscape. We consider some of the general material culture of water in Niamey, then we explore the trends and impacts of plastic packaging of water on perceptions of water quality. Next, we look at the advertising and branding of water by analyzing highway billboards with water-related themes. Finally, we return to our discussion of sachets begun in the previous chapter, but this time we specifically analyze the branding of sachets and consider the lack of brand loyalty among consumers in Niamey.

Material Culture of Water in Niamey

If water's materiality, as defined by Orlove and Caton (2010: 403) and in Chapter 1 of this book, is "the physical attributes of water that affect its relation to the human body and environment that shape its use," then water's materiality includes the objects and systems that facilitate these relations between water and humans. Water's materiality also includes the objects and systems that facilitate its commoditization. Thus, "materiality" is not just tangible things but also policies and practices that commoditize water, govern access, and connect people with their environments.

We have explored some of water's materiality in this book already. Chapters 4 and 5, on *ga'ruwa* water delivery systems and the sachet water economy in Niamey, consider the materials involved in formal systems of water delivery, like large water purification plants, the technologies associated with them, and the networks of water mains, pumps, and smaller pipes that transport purified water through many parts of the city. *Ga'ruwa* delivery and sachet water are both hybrid systems of water provision because they exist to fill the gaps in water access that plague Niamey, but at the same time, both depend on the piped network carrying purified water to public standpipes (in the case of *ga'ruwa*) or private household connections (in the case of sachet water).

Both *ga'ruwa* delivery and sachet water have their own unique material cultures as well. *Ga'ruwa* use metal pushcarts made by local blacksmiths, they reuse cooking oil containers to transport and deliver water, they use plastic funnels to facilitate the placement of water into private storage containers in households and businesses, and they decorate their carts using flags from their country of origin, plastic flowers, or other distinguishing markers of personal and national identity (Figures 6.1 and 6.2). Sachet water, a more mechanized process, involves materials that have traveled through much longer commodity chains, like the machines that produce sachets and the large rolls of plastic that ultimately form the sachet upon production. The coolers and plastic bowls used by mobile vendors to keep the sachets cold as vendors walk city streets are part of the consumer goods trade between Niger and China (and other industrialized countries). Some sachet water vendors use small metal pushcarts—smaller than those used by *ga'ruwa*—to transport their products.

Of course, materials carry with them meaning, which we explored in Chapter 1 and, in relation to sachet water, returned to in Chapter 5. For example, the machines that help create the sachets mean economic opportunity for their owners but, at the same time, usually significant debt, as substantial loans and credit had to be obtained, and time invested, to secure the machines in the first place. The home connection to formal piped water and electrical networks symbolizes modernity

Figure 6.1 *Ga'ruwa* cart with yellow plastic jugs. Photo by the authors.

Figure 6.2 Decorations on a *ga'ruwa* cart. Photo by the authors.

while at the same time facilitating a hybrid economy that characterizes marginalized areas and people who often are underemployed and lack access to wage labor opportunities.

Material cultures are not fixed, however; they change over time as new uses are discovered, new technologies are created, new adaptations are required, and new alternatives to traditional materials are introduced. We observed this most strikingly in water storage systems in Nigérien homes. While shadowing water vendors, we were often permitted to enter the compounds where a *ga'ruwa* was delivering water and noted how the relatively large quantities of water purchased by the households was stored long term. Traditionally in Niger, large clay pots made in the village of Boubon, about twenty miles outside of Niamey, were partially submerged in the ground to naturally cool the water contained within them. They were covered with woven mats, called *faifai* in Hausa, to keep out dust and insects, and tin cups were used to extract water from the pot for drinking or cooking. In our fieldwork, however, we observed more often the use of large plastic containers imported from China for long-term storage of water (Figure 6.3).

Not only were the impacts of this material transition concerning for their implications on the local economies of clay pots and *faifai*, but so were the dangers from chemical leaching into drinking water stored for long periods of time in plastic containers, often in areas of the compound subject to high temperatures. Some interviewees acknowledged the potential harmful impact of storing water in plastic,

Figure 6.3 Plastic containers next to clay pots (from Hassane's compound). Photo by the authors.

others explained that after a while, the water in the plastic containers would taste bad, and still others indicated that the relatively low cost of plastic containers compared to clay pots made them an obvious choice in situations where household income was tenuous. Other traditional elements of water's material culture have changed as well. Goatskin bags and tire tubes were commonly used until about twenty years ago for hauling water from wells on the urban periphery and in villages. Tin petroleum containers were used to store water. These have now been partially or completely replaced by plastic materials.

Some of the materials used in long-term water storage are changing as well, while others are not. Throughout Niamey, large cement water towers (called *chateaux* in Niamey) can be seen standing high above the homes in several neighborhoods (Figure 6.4). These water towers are owned and managed by Veolia-SEEN and are visible indications that certain neighborhoods experience water shortages during the year. In the event of long-term interruptions in water and/or electric services, water is released from these towers, and gravity carries the water to public standpipes and private taps via the piped network. The closer people live to a neighborhood water tower, the more likely they are to get water when it is released, and the higher the water pressure. Those living farthest from the towers may not actually get water if much of it is taken by households closer to the towers. These towers are an extension of the piped network, which does not always deliver water to all parts of the city it is designed to serve. Their permanence and importance in water delivery is evidenced by the fact that several neighborhoods share their name with the number assigned to the water tower (the neighborhood Chateau 1 is an example).

Although the materials used to construct the large neighborhood water towers have not changed significantly, the availability of cheap plastic imports from China has changed the nature of long-term water storage. Instead of long-term water storage, and planning for water service interruptions, being solely the responsibility of the city government, the introduction of cheap plastic imports has made long-term water storage an option for individuals and small groups. Residents of Niamey can now store water long term in their compounds using large plastic containers imported from China and designed for this purpose. If possible, the container is elevated to take advantage of gravity in the event water needs to be drawn. We observed these private water towers in both wealthy (Figure 6.5) and poor (Figure 6.6) neighborhoods in Niamey, which indicated several things. First, the wealthy neighborhoods are not immune to water service interruptions,

Figure 6.4 Large neighborhood water tower (called a *chateau*). Photo by the authors.

although our interviews indicated that those interruptions are less frequent in wealthy neighborhoods. Second, we observed that the large private water containers in wealthy neighborhoods get emptied and refilled every few weeks to reduce the impact of leachates and other contaminants in the water. The wealthy, who pay less for water than the poor, can afford to essentially waste water in this way, as the compounds we observed simply let the water run onto the ground as it was emptied from the container.

We also observed these water towers in poor neighborhoods. In one neighborhood, for example, several households collectively purchased a large plastic storage container and the materials to elevate it. In the

Figure 6.5 Private water tower in a wealthy neighborhood compound. Photo by the authors.

event of water service interruptions, the households collectively use the stored water until service is restored or the water runs out. However, our interviews indicated that the water stored in elevated plastic containers in poor neighborhoods was changed much less frequently than that in wealthy neighborhoods.

Figure 6.6 Community private water tower in a poor neighborhood. Photo by the authors.

Though not common in Niamey as discussed in Chapter 3, private wells also offer a long-term water storage option and come with their own material culture, including digging tools, mud bricks, ropes, and the recent additions of pulleys and plastic buckets. Typically found on the city's periphery, material investment comes upfront, and theoretically only minor maintenance is involved afterward. If water is used sustainably, and if sufficient amounts of rainfall are received, a private well can serve as an inexpensive long-term storage method for water.

Boreholes are also changing the material culture of water access. Although boreholes are becoming more common in villages in the Sahel, there are some in Niamey as well. We mentioned one such urban borehole in Chapters 1 and 3 when we described the charity actions of a Saudi Arabian-based Islamic organization. In addition to the actual materials that encompass the borehole, an extensive mix of technology and material culture enables their creation. The subterranean water resources in Niger are increasingly seen as the "solution" to some of Niger's water availability issues, evidenced by the increasing number of businesses related to borehole drilling (Figure 6.7). Boreholes mean more efficient water access, at least while supplies last, and in the example of the Islamic organization's borehole in the Abidjan neighborhood in Niamey, the water is free. (In contrast, in the village of Matankari, where we studied water access, those taking water from the borehole-fed water taps had to pay for it, while water from the village wells was free.) However, there are also water quality concerns for urban boreholes, as subterranean water sources that lie beneath cities can easily become contaminated.

Figure 6.7 Advertisement for a borehole drilling company. Photo by the authors.

Although our focus in this book is on hybrid drinking water economies, we cannot leave a discussion of the material culture of water without some commentary on the materials involved in more extravagant and luxurious water uses. These are, of course, found exclusively in wealthy neighborhoods (or in wealthy compounds located in poor neighborhoods), as they fall outside the category of materials used to facilitate access to drinking water. For example, our house guards were determined to wash our twenty-year-old Toyota Corolla daily. This was partly due to the fact that they seemed bored—they never had to confront an intruder during our stay—but more importantly because they assumed that as relatively wealthy people we expected this service even though we never directly asked for it. We felt guilty about squandering water in the Sahel and eventually told the guards that washing the car once every ten days was quite sufficient. Even though Niamey, with its location on the Niger River, has more green vegetation than is found in most of the rest of the country, it is still striking to us when we see lawns and pools in people's homes or in businesses.

The lawn at the American International School of Niamey, for example, is watered regularly (Figure 6.8), as is the lawn surrounding the US ambassador's residence (Figure 6.9). Hoses, spigots, sprinkler systems, and electricity facilitate the use of immense amounts of water to maintain this element of material culture in Niger. Many compounds in wealthy neighborhoods also have pools, and in our search for housing in Niamey during our Fulbright sabbatical leave year, our desire to avoid a pool made the search for housing more difficult, even though, at the same time, it also reduced the rent price of the homes we considered. What makes the presence of lawns and pools even more striking, though, is that the water used to maintain them is the same water that people drink—chemically purified water from the piped network. As we consider the water gap between the rich and the poor, we have to consider that the extravagant use of water by the wealthy drives up the cost of water overall, as lawn water comes from the same source.

Thus, materials used to access, transport, and store water carry with them messages about socioeconomic class, perceptions of quality, and household priorities. They are elements of water's materiality and are part of the process by which people assign value and meaning to water. In essence, they are part of water's social life. We now focus on one specific type of plastic material that shapes perceptions about water: plastic packaging.

Figure 6.8 Lawn and sprinkler at the American International School of Niamey. Photo by the authors.

Water and Plastic Packaging

Plastic plays a key role in the materiality of water. In their book *Plastic Water* (2015), Gail Hawkins et al. explore how plastic, particularly in the form of bottles, created new drinking and waste realities. Although water was marketed long before the invention of plastic, it was plastic packaging, especially in the form of the polyethylene terephthalate (PET) bottle (and, we argue, plastic sachets) that moved commoditized water into the category of "fast-moving consumer good" (Hawkins et al. 2015). Plastic water containers have different meanings in different settings and for different people, which is part of the reason water has such a complex "social life" (Wagner 2013). According to Hawkins

Figure 6.9 Lawn at the US ambassador's residence overlooking the Niger River. Photo by the authors.

et al. (2015: xiv), the PET bottle can "exist as a product, as a personal health resource, as an object of boycotts, as part of accumulating waste matter," and, we would add, as a valuable resource in its recycled-reusable form, as a surface ideal for marketing, as a generator of fossil fuels and a consumer of petroleum in its production, as a symbol of corporate hegemony, as a method of storage, and as an indication of material wealth.

There is a market for bottled water in Niamey, but its consumption is relegated to wealthy Nigériens and the expatriate community, as they are the only people who can afford to purchase it. Six Nigérien bottled water companies dominate the market in Niamey: Belvie, Dallol, Diago, Rharous, Tasnim, and Telwa. Jirma, produced in Burkina Faso, is also widely available. In addition, four French brands—Eau Vitale, Evian, Fifa, and Perrier—are available in the most upscale supermarkets and restaurants of Niamey. The West African–produced brands are about 25 percent cheaper than the French imports, but all are extremely expensive relative to other forms of commercial drinking water. Bottled water on average is more than 800 times as expensive as water taken directly from public standpipes.

When Keough arrived in Niger in 2016 as a Fulbright scholar, she underwent a briefing at the US Embassy in Niamey, which included a meeting with the director of health services for the Embassy. She handed Keough a laminated piece of paper about the size of a business card which listed the six main bottled water brands, all of which the US Embassy tests for quality. According to the Embassy's report, the three most popular West African brands all failed the Embassy's water quality tests, while the three French brands passed. Furthermore, the director told Keough that once a bottled water brand fails a water quality test, it is never retested; it is forever placed on the "failed" list (complete with red ink indicators). Interestingly, the Embassy also periodically tests the water quality from the piped network, and according to the health official with whom Keough spoke, the piped water in Niamey has never failed the Embassy's water quality tests.

Another concern about plastic packaging involves the landscapes of waste and the environmental pollution it creates. In the description of the "solutions" to the Flint water crisis (from Chapter 1), we considered the environmental impact of supplying an entire city with bottled water, as an alternative to toxic piped water, for several years. Michigan cities do not have strong recycling programs, and there is not an inherent culture of recycling in Michigan cities, compared to other regions of the US like Seattle. Our city, Saginaw, for example, only instituted curbside recycling pickup seven years ago. Prior to that, residents had to pay a private company to come to collect materials for recycling or drive to the remotely located recycling center themselves with their recycled household products. Few people did this.

In Niamey, PET bottles do not constitute a large portion of material waste deposits because they can be reused and repurposed for long periods of time after the initial consumption of their contents. More ubiquitous, however, are sachets made of polypropylene plastic, which are consumed across the socioeconomic spectrum. They are produced in much larger quantities than plastic bottles, and, at least at the time of our research, there appeared to be no apparent reuse or recycling of sachets after the water they contained was consumed. The commodity and value chain appears to end after the consumers discard the sachets. The discarded sachets, as we indicated in Chapter 5, clog sewer drainage canals, many of which remain open alongside streets, and many end up in the Niger River, affecting other environmental and biological systems.

Plastic packaging also creates environmental degradation in its production. As a petroleum-based product, the global-scale transportation and consumption of plastic goods increases pressure on petroleum resources, driving demand and prices, instigating extraction in harmful and unsustainable ways, and contributing to greenhouse gasses in the atmosphere through the emission of fossil fuels during the transport of plastic materials. Harmful chemicals and toxins pollute the atmosphere both in the process of extracting petroleum and in factories' production of the plastic materials. These realities have created what Hawkins et al. (2015: 147) call a "vilification" of bottled water. In this way, consumption in Niger is linked to larger, global environmental problems. Nigériens do not directly produce the greenhouse gasses responsible for global warming, but increased consumption of plastic materials indirectly contributes to the climate crisis.

Although approximately 95 percent (we estimate) of the sachets sold in Niamey are produced through the cottage industry we described in Chapter 5, there are three sachet labels that are produced by water companies on an industrial scale: Amico produced by Souley Group S.A., Mataba produced by Mataba Torodi International, and Zam-Zam produced by Niger Lait S.A. (Niger Lait also produces sachets containing milk and liquid yogurt). We had the opportunity to tour one commercial-scale sachet production facility in Niamey, Niger Lait, and interview the plant manager. Commercially produced sachets contain chemically purified water produced through the process of reversed osmosis, similar to the way bottled water is produced. In the Niger Lait plant, the chemically purified water was sent through a system of pipes to the automated machines. These looked almost exactly like the ones we observed in private homes in poor neighborhoods, except in this case, four machines (each with one employee) were operating simultaneously, thus increasing exponentially the number of sachets produced at one time. The workers were wearing protective shoe, hair, and hand covering to limit opportunities for contamination, as were we and our guides while on the plant floor.

We were particularly curious about why these sachets, made of higher quality plastic and more professionally designed colored labels compared to the cottage industry sachets, are not more visibly present in Niamey's urban landscape (Figure 6.10). Our discussion with Niger Lait representatives, however, revealed a product hierarchy in sachet water. Sachets produced through industrial methods are more than twice as expensive as those produced through the cottage industry. The location of the Niger Lait plant, in Niamey's industrial zone, makes it largely inaccessible to mobile vendors, most of whom are of a socioeconomic class that cannot afford motorized transportation. According to the Niger Lait plant manager, there is no local distribution of water sachets. Most of the sachets produced at the plant are purchased by importers from Burkina Faso, who cross the border, load their trucks at the plant, and resell the sachets in Burkina Faso. Wealthy Nigériens

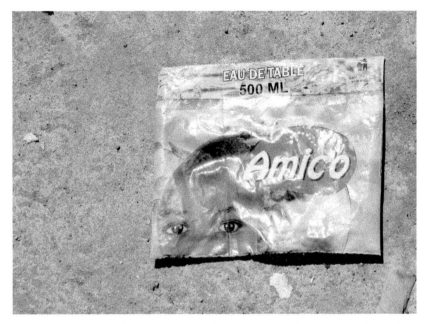

Figure 6.10 Picture of Amico water sachet, commercially produced. Photo by the authors.

in Niamey also drive directly to the plant and purchase the sachets in bulk.

We discovered this after being invited to a naming ceremony, where we were served commercially produced sachet water, and we subsequently asked our host where he had purchased it, since we had never seen it for sale on the street. In further contrast to the sachet cottage industry, the commercially produced sachets are not sold cold, even though they are more expensive. It is assumed that the buyer has access to refrigeration. Moreover, although Niger Lait has several billboard advertisements for its products throughout the city, none of them advertises sachet water (a few signs advertise the plastic cups of water one might receive as part of an inflight meal on an airplane). Despite the fact that the product falls into the same category, large-scale industrial production of sachets is not displacing the ones produced in-house.

In its current form, there does not seem to be direct competition between commercially produced sachets and cottage industry–produced sachets. However, at least among Nigériens who can afford it, the commercially produced sachets are more desirable, not only for their appearance and perceived quality of water contained in them but also as a status symbol used to indicate the relative wealth of the host and appreciation for guests during special occasions.

Thus, as water has become a commodity, often controlled by the private sector and with profit-boosting potential, its material culture has expanded to include ways that it is branded and promoted through advertising. The next section explores these phenomena.

Advertising and Branding Water

In Niamey, most of the advertising of water occurs along major highways and city streets in the form of large billboards. These billboards combine words and images to communicate messages to broad audiences and to promote brands, but these brands are not simply product representations: they are mediums that influence and shape relations and actions between products and consumers (Lury 2004). In this way, public space along Niamey roads becomes symbolically constructed through the use of billboard advertising. Semiotic approaches to material culture encourage us to consider what messages objects send and how objects may stand for something else (Berger 2014).

Billboard advertisements for water are literally signs (a message to an audience), but they also represent the hegemony of corporate control of water in Niamey, and they present juxtapositions between promoting consumption (which produces profit) and promoting conservation of a valuable resource. Billboards are examples of visual imagery that can have powerful and emotive effects on the viewer (Joffe 2008) and can potentially influence public engagement in an issue (DiFrancesco and Young 2010). These billboards contain words and are thus part of the urban linguistic landscape (Calvet 1994). In Niamey, however, the images used are of equal or greater importance to the words because Niger's adult literacy rate is so low (19.1 percent, according to the United Nations Human Development Reports 2016).

In our analysis of the city's streetscape, we observed two categories of billboards associated with water along main thoroughfares: signs promoting the commercial sale of water products, bottles, and cups (see Figures 6.11 and 6.12) and public service announcements urging consumers to conserve water (see Figures 6.13 and 6.14). We explore themes in water advertising using these four billboards as examples, and then we move on to explore branding by considering labels on sachets sold in Niamey.

Figure 6.11 Billboard advertisement for Belvie bottled water. Photo by the authors.

Figure 6.12 Billboard advertisement for cups of water. Photo by the authors.

Figure 6.13 Billboard advertisement for water conservation (2015). Photo by the authors.

Figure 6.14 Billboard advertisement for water conservation (2016). Photo by the authors.

Bottles and Cups

In an attempt to maximize profits, corporations typically try to assemble a diverse market—they develop a product that will appeal to a diverse audience and market it to as many different social groups as possible. As Hawkins et al. (2015) explain, branding is central to this goal of market diversification. They argue that the brand has "prompted significant transformations in the wider valuations of all drinking water" (Hawkins et al. 2015: 28). Certainly in the United States, many consumers view bottled water as safer than tap water (Annin 2006), despite the fact that bottled water is tested less regularly than municipal tap water and many bottled water companies purchase water from municipal water supply systems. In Flint, Michigan, bottled water was the solution to a contaminated municipal water supply, and in the immediate term, the safe option.

In Niger, the consumer base for bottled water is much smaller than it is in wealthy countries. The cost is out of reach for most of the population. Some bottled water is imported, which drives up the cost as well, since Niger's landlocked position increases the transportation costs of goods. One local brand of bottled water is Belvie, a label produced in Niamey by an Indian entrepreneur who came to Niger, saw a niche that he thought he could fill, and built and opened the Belvie water bottling plant. Belvie uses water from an underground aquifer on the urban fringe and is able to undersell imported bottled water brands (Probyn 2016). Belvie is the most common bottled water brand advertised on highway billboards (Figure 6.11).

The Belvie billboard advertisement clearly indicates its target audience. Pictured are four adults and one child, all wearing either clothing to indicate their professions (the doctor and the athlete) or expensive African prints—clearly they are part of the wealthy class. The company's name is a contraction of the French words *belle* "good" and *vie* "life," and by implication consumers can access the good life through consuming their product.

The company's motto, "La Pureté à chaque goutte" (Purity in each swallow), promotes the idea that their clear water in plastic packaging is of higher quality than the tap water in Niamey's municipal system, especially with the added phrase "Eau Minérale Naturelle" (Natural Mineral Water). The choice of adults is clear: the athlete who relies on water for successful performance and the doctor who should know what is best for health both endorse the water. Furthermore, it is suggested that Belvie water is best for children, both born and unborn, as the woman pictured in the advertisement is pregnant. The child and the athlete are both in the process of drinking the bottled water, indicating the water's availability for immediate consumption. Finally, the company's business certification number is included at the bottom left of the advertisement, which encourages the viewer to consider that the company's claims are valid.

The billboard advertisement for Zam-Zam cups of water (Figure 6.12) takes a somewhat different approach to the idea of water purity, one that is subtler. The Niamey-based company Niger Lait, which also produces sachet water, produces the Zam-Zam label. (Another Zam-Zam brand of sachet water is made in Dogondoutchi, Niger.) We discussed the cultural symbolism of marketing this brand named after the Zamzam Well in Mecca, which is reputed to contain holy or healing water in Chapter 1. The fact that plastic cups of water were even the subject of highway billboard advertisements in the first place was surprising, given that the only place we saw these for sale (of any brand) was at the beverage stand at one of the large private health clinics of Niamey. Other Niger Lait products, like sachet milk and yogurt, are available at roadside boutiques and in supermarkets in Niamey, but outside the clinic, the only place where we saw cups of water available for consumption was as part of our meals on our Air France flight (and these were not the Niger Lait brand). (At least one other brand—Akakos—of water in cups is produced in Niamey.)

The billboard itself has some striking discontinuities. The subtext in the Zam-Zam ad reads, "une eau fraîcheur" ("refreshing water," although *fraîcheur* can also mean "coolness" and "youthfulness," and Zam-Zam is an implied claim of purity), but no claims about purity appear on the sign. This is interesting because the Niger River is fresh water, but no one would make any claims about its purity. In another example, although a Nigérien company produces the product locally, the family of four depicted in the advertisement looks like the stereotypical wealthy modern West African family: a small family with only two children, who obviously do not need to be concerned about water conservation, as the water is shown splashing around the image. Furthermore, Niger Lait is pitching their product to a particular modern class of people, including those who may have experience drinking water from these types of cups on airplanes. Unlike the Belvie advertisement, no one in the Zam-Zam image is wearing African cloth. All don secular Western-style clothing. Even more striking is that the woman (presumably the mother of the two children) is wearing a skirt that falls above the knee, a style that is very unusual in Niger.

Another difference is in the depiction of the water itself. In the Belvie advertisement, all the water is contained within the plastic packaging, and in the Zam-Zam advertisement, the water is splashing out of the packaging, and over the family, who appears excited about these conditions. Perhaps the splashing symbolizes the position wealthy people are in that allows them to play with water. No one in the advertisement is depicted consuming this water, although it appears the girl is reaching for one of the plastic cups. Despite the fact that a local company produces Zam-Zam, no government certification number is included; however, the locations of the production facility as well as local phone numbers are provided. In some ways, the business address and phone numbers, as well as the reference to Zam-Zam in a predominantly Muslim country like Niger, makes the Zam-Zam label appear more local than Belvie, which has no specific references to Niger in its advertisement. In other words, in the Belvie advertisement, the people appear more local, but in the Zam-Zam advertisement, it is the text that situates the company solidly in the *umma* (global Muslim community) in general and in Niamey in particular.

Water Conservation

We first noticed public service announcements in the form of billboards promoting water conservation during our 2015 fieldwork in Niamey. The two billboards pictured here are the only two that appeared during our research trips, although several copies of the same billboard were found throughout the city, at least on the left bank. Both are published by Veolia-SEEN and carry a very similar general message: *"L'eau, un bien précieux. Economisons-là! CHAQUE GOUTTE COMPTE"* (Water is a precious good. Conserve it! EVERY DROP COUNTS). Presented against the broader trend of water privatization and commoditization, however, these messages carry other meanings and are even oxymoronic.

The subtext of Figure 6.14, for example, includes the phrase *"une eau sain partout and pour tous"* (clean water everywhere and for everyone), a phrase that aligns well with water-related sustainable development goals but that seem hypocritical considering the for-profit practices of Veolia-SEEN in Niamey and elsewhere that are anything but pro-poor. Indeed, the irony is that Veolia-SEEN wants all Nigériens to purchase as much of their water as possible to maximize their revenues. They are, however, willing to settle for what they can receive from Nigériens who can afford their product; otherwise, Veolia would not have remained in Niger for almost two decades.

Other juxtapositions are evident as well. For example, the SEEN logo depicts a pastoral nomadic Fulani woman—identified by her hoop earrings—carrying a calabash of water atop her head. Thus, the SEEN logo references a traditional Nigérien culture, although a Fulani woman is one of the least likely consumers of their product. Furthermore, both advertisements are promoting water conservation, and both depict leaking faucets, one of the problems plaguing parts of Niamey's aging piped system. Second, in the advertisement in Figure 6.13, drops of water are falling onto a person's hands, and presumably the individual will drink this water immediately afterward. Not only is this an inefficient method by which to consume water, it is also less sanitary, although in Niger, Muslims perform ablutions, involving rinsing their mouths with water using their hands.

Furthermore, in Figure 6.13, only the hands are visible. In Figure 6.14, the child's body from the torso up is depicted. He uses a bowl to catch the drips from the faucet, an action supporting the subtext of the sign, "L'eau c'est la vie, évitons de la gaspiller" (Water is life, let's not waste it). A grateful smile appears on his face. Earlier we mentioned the limited impact that written words in advertisements have on a large portion of the population due to low literacy rates, but this second advertisement offers an active visual solution that supports the message. The child catches water from the leaking faucet in a small plastic bowl, the kind of inexpensive plastic imported good easily accessible throughout the city. Even without the written message about water conservation, a viewer could be inspired to copy the actions of the child in an effort to collect as much water from available faucets as possible.

Taken collectively, billboard advertising of water in Niamey contrasts consumption with conservation, but it also can be viewed through the three modes of persuasion (ethos, pathos, and logos) common to the discipline of rhetoric. All four billboards contain elements of ethos, where references to authority are included. Belvie includes the product certification number, Zam-Zam states the physical address (and phone numbers and websites) where production occurs, and the two water conservation billboards contain the SEEN and Veolia logos. The billboards also contain examples of pathos, or emotional appeals. The people in the Belvie and Zam-Zam advertisements are happy—clearly life is better with plastic-packaged water. Figure 6.14, with the child holding water under a dripping faucet, elicits an emotion in the viewer that not only is conservation good for the environment, but we must do this for the (poor) children. Figure 6.14 is probably the best example of logos, or attempts to persuade an audience using logic or reason, where a specific water conservation technique is depicted so that even someone who cannot read the words might be inspired to conserve water in this way. The Belvie advertisement depicts the consumption of water—it is not about water conservation—and communicates that good health can be obtained by drinking their bottled water. The presence of the doctor in the advertisement reinforces this idea. (The doctor is an example of ethos, adding authority to the perceived purity and health benefits of drinking bottled water.) Juxtaposed to Figure 6.14 is the Zam-Zam advertisement, where water is flung about the image in a careless way.

The two types of water billboards obviously have different objectives. Belvie and Niger Lait are selling products by associating them with a lifestyle—a happy, modern, Westernized, small family lifestyle. In contrast, the Veolia-SEEN billboards are really public service announcements, even though Veolia-SEEN is a for-profit entity. Their billboards are designed to create a good humanitarian image. The Niger Lait advertisement, where water is flung carelessly about the poster, stands in contrast to the concerns over water we heard at all levels of society. No direct references to Islam, the Niger River, or local languages appear on the signs (Zam-Zam is an indirect reference to Islam), despite the fact that Belvie and Niger Lait are Niamey-based corporations and SEEN is a Nigérien public entity in partnership with Veolia. While more local references and audience-appropriate means of communicating ideas might have a greater impact, water-related signs in both categories (commercial advertisements and public service announcements) are in the minority on Niamey's billboard advertising landscape, so it is difficult to determine if these signs have a significant impact on their intended audiences.

Branding Sachet Water

Branding through labeling has been the focus of several studies on bottled water (Collins and Wright 2014; Hawkins et al. 2015; Wilk 2006), but we did not find any publications on branding in the sachet water industry. During field research trips in 2015, 2016, and 2017, we collected sixty-five different sachet labels that had been discarded on roadsides. We know that there are many more brands, but we were unwilling to pick up particularly dirty bags and we think that sixty-five is enough to provide a representative sample. The labels share some elements despite their diversity. All—with the exception of one commercially produced brand—use blue ink on clear plastic to include a name, indication of the place of manufacturing, and the quantity of contents—usually fifty centiliters but occasionally sixty centiliters. Most include a logo or imagery and at least one expression. These labels reveal much about the creativity of those who make them but not very much about consumers, as we will explain below.

English is the most commonly used language on the labels that we collected in Niamey, followed by French, Arabic spelled out in Latin script, and Hausa. For unknown reasons, we found none with Zarma, even though Zarma participate in the *piya wata* economy about as much as the Hausa do. English is used exclusively on the bags manufactured in Nigeria. Most of the bags made in Niger include both English and French, and some use English only. Of course, claims made on labels are just that, claims. For example, a man working in Nigeria with a blue ink stamp can use a Niamey address on his sachets. Furthermore, there can be a difference between where the bag was stamped and where it was filled. We did not find any evidence that Niger is importing prefilled sachets from water from Nigeria.

Nigériens now have their own automated sachet filling machines, and transporting water is very expensive due to its weight and customs fees. Thus, we conclude that Nigériens are importing labeled plastic sachets from Nigeria and later filling them in Niger. This also means the Nigerians are deeply involved in the international plastic bag trade, including some who are working as producers in Niamey, as discussed in Chapter 5. The most common types of names are corporate-sounding names or personal names, such as Dontex, JC & JC, MDY, and Saley Dogo. A few used religious names, such as Showers of Blessing, Halal, and Zam-Zam. Hausa is used in the brand names on only five bags.

Among the sixty-five bags in our sample, thirty-six indicate that they were manufactured in Nigeria, twenty-four in Niger, two in Benin, one in Côte d'Ivoire, and two that did not indicate a place. The Nigerian bags claim origins primarily in a variety of places in Northern Nigeria. Two-thirds (sixteen) of the Nigérien bags indicate that they were printed in Niamey, whereas one-third (eight) indicate that they were made elsewhere in Niger, including four in Tahoua and one each in Ballaiyara [sic], Gaya, Maradi, and Torodi. Clearly the bags have social lives that involve transport over great distances, supporting Stoler's (2017: 1) conclusion that "sachet water has become a multibillion-dollar industry—and a veritable consumer phenomenon—throughout West Africa."

Water imagery is the most widely used motif on the sachets, especially depictions of Saharan oases with palm trees. Other imagery in the sample includes waterfalls, waves, water droplets, icebergs, and fish (that live in water). Islamic imagery was almost completely absent; however, the Halal brand made in Niger and the Dan-SADA brand made in Sokoto State, Nigeria feature the Islamic crescent moon and planet. The KHALIFAT label made in Ibadan, Nigeria indicates that their bags contain "prayer water." Only the Zam-Zam brand manufactured in Dogondoutchi, Niger includes any Hausa wording: "Tsab-tac-cin Ruwan Sha" (Clean Drinking Water). In this case, the brand name itself implies religious purity. All Nigerian brands are labeled as "table water," whereas the Nigérien brands are typically labeled as "pure eau potable" or simply "eau potable." Many English- and French-language bags use the ubiquitous expression "Water is life" and "L'Eau est vie."

Virtually all of the Nigerian labels include the words "NAFDAC Reg No." followed by a five- to seven-digit number. NAFDAC is an acronym for Nigeria's National Agency for Food and Drug Administration and Control, which is charged with monitoring the cleanliness of sachet water production. The NAFDAC label contributes to the ethos or credibility of the product. Niger lacks an analogous institution. However, as indicated earlier, just because a bag includes the words "NAFDAC Reg No." and a number does not guarantee that NAFDAC actually regularly inspects the production facilities or that the bags were even filled in Nigeria. Furthermore, many bags include the words "NAFDAC Reg No." but lack numbers. And, as indicated earlier, independent testing of sachet water in Nigeria regularly finds that much of it is contaminated. Finally, most bags include the words "manufactured on" or "best before," but none include dates. Most bags use an outline image of a person depositing a bag in a trashcan despite the reality that trashcans are rare in the region.

Clearly the people responsible for branding use a lot of creativity in selecting names, logos, imagery, and words for sachet water. We remain puzzled by this level of effort, as indicated at the beginning of our discussion of the branding of sachet water, because consumers in Niamey do not seem to care much about it. We neither observed consumers examining labels nor asking for brands by name. Rather, consumers simply use the generic term, *piya wata*. None of the seventy consumers that we interviewed indicated a brand loyalty. However, some *piya wata* vendors indicated loyalties with particular producers that they know in their neighborhoods, but these producers regularly change the bags that they use depending on what empty bags they can find for sale in the markets of Niamey. Customers in Niamey care only to see that the water appears clear, is machine sealed, feels cold, and has a label. Any label suffices, because labels distinguish contemporary *piya wata* from the previous practice of selling manually tied, unlabeled bags of water. It is possible, however, that labeling implies a desire to placate particular audiences, such as NGOs, government ministries, and expatriates.

In these brands and billboards, water appears as a metaphor for life and for profit, whether it be through consumption or conservation. Water is life because it must be literally consumed by living individuals, such as the athlete drinking water in the Belvie ad, and conserved as a valuable commodity, especially in the Sahel. Water is also profit, as consumption generates income for individual sachet vendors, utility companies, and packaging corporations. Branding is an attempt to increase a product's consumer base and maximize profit, but we see little impact from these efforts in Niamey. Finally, water's materiality underlies and provides cultural meaning for the hybrid systems of delivery we discussed earlier in the book. Underlying this materiality is the phenomenon of global plastic capitalism, which provides inexpensive, often low-quality consumer goods to countries like Niger, whose manufacturing base is not well developed.

While we as researchers observe the negative impacts of plastic materials on the environment in Niger, there appears to be little effort in Niamey to combat this problem. The black plastic bags used in market transactions continue to represent a finalized economic transaction and are discarded, are blown with the wind, and they collect on bushes and trees or street corners. No apparent recycling of water sachets occurs, at least as of the time of this writing. And these local practices and systems are, of course, connected to larger patterns of power, governance, and access to natural and material resources.

Conclusion

Youssoufou (a pseudonym) and Adamou, his youngest brother, reside in a humble *banco* compound in Zongo that was given to him thirty years ago, when his father retired and moved to a small town in south central Niger. (Zongo is an old neighborhood in the center of the city surrounded by state office buildings, a city hall, and two markets.) Youssoufou has found only occasional work—roughly one day per month—as an electrician, mechanic, and tutor over the past three decades. He has survived into his sixties with relatively good health because Zeinabou (a pseudonym), his wife, shared her earnings with him, and due to the luxury of living rent-free and the generosity of family and friends. Zeinabou moved into the compound nearly twenty years ago, when she and Youssoufou married. She sold meals on the street just a few blocks away. Zeinabou gave this up ten years ago, when her arthritis worsened and she fell into debt due to offering credit to too many customers who never made good on their debts. Adamou, now in his forties, has not worked a day in his life.

Faced with the loss of Zeinabou's small but steady income, Youssoufou and Zeinabou decided that they needed to do something. They noticed the rising popularity of machine-sealed sachet water and figured they could enter the business and make money relatively easily. They also knew that many of the fishmongers at the nearby Petit Marché did not own refrigerators or freezers and purchased blocks of ice daily. Using much of their savings and gifts from family and friends, they purchased three secondhand refrigerators with freezers. They could not afford to buy a sachet filling and sealing machine, so they had to settle for a simple, electric manually operated device.

Youssoufou and Zeinabou were able to generate some income for a couple of years, particularly during the hot seasons. (They used unfiltered *ga'ruwa*-delivered water since they are too poor to have a tap in their home.) However, they faced many challenges. They did not have young children, so they had to hire children from the neighborhood to sell the water on the street. Their monthly electricity bills skyrocketed despite frequent, long power outages that warmed their water and melted their ice. Furthermore, Youssoufou found his manual sealing device tedious and annoying, as he could fill and seal only about one bag per minute and he frequently suffered painful electrical shocks.

Disaster struck in 2012, when the Petit Marché was destroyed by fire. Youssoufou and Zeinabou made more profit selling ice than sachet water, but the fire scattered the fishmongers to markets too distant for them to access. This not only put them out of business but also significantly reduced the quality of life in Zongo. Many vendors who had worked in the Petit Marché simply moved to the streets of Zongo, where there are no public toilets. The storm sewer on the small street where Youssoufou and Zeinabou live in the shadow of a high-rise government ministry building has been clogged—primarily with sachet water bags and black plastic shopping bags—since 2012. A forty-meter section of the road is now a fetid cesspool. Market vendors and a team of *ga'ruwa* who sleep on the street just twenty meters from the cesspool defecate and urinate on the street all day long just ten meters from Youssoufou's compound. Youssoufou and Zeinabou now endure a lot of sleepless nights worrying about how they will survive selling just a few small, hand-tied bags of ice a day to neighbors, concerned that their water is no longer safe, inhaling toxic fumes, and disturbed by the growing numbers of mosquitoes and biting flies brought by the cesspool. They are angry too. Youssoufou regularly points out that they live just two blocks from city hall and that government workers in their air-conditioned offices see the cesspool every day but have no concern for the poor residents of Zongo.

Meanwhile, the city has neither rebuilt the Petit Marché nor allowed vendors to use the space. It simply sits there empty, surrounded by a cheap tin fence. Most Niameyans cannot fathom why a whole square block in the city center has gone unused for six years as of the time of this writing. However, some people believe that President Mahamadou Issoufou wants to use the land as part of his urban development plan, "Niamey Nyala." A few conspiracy theorists think that the state torched the market for this purpose. Furthermore, for several years rumors have circulated that all of Zongo will be razed and replaced by government buildings and foreign businesses as part of Niamey Nyala.

Niamey Nyala is President Issoufou's plan to renovate and modernize the appearance of Niamey (République du Niger 2012). *Nyala* is a Zarma term meaning "sparkling" or "modern." Some Niamey Nyala advertisements also use the French term *"Niamey La coquette,"* which means "charming" or "attractive." Niamey Nyala projects were initiated several years ago and are scheduled to be completed in 2020, just before Issoufou completes his second and final term as president. Most

Nigériens know very little about Niamey Nyala because state press releases have been released only sporadically and are often vague. Those who have heard about it tend to be highly critical, especially if they are poor.

Niamey Nyala includes four primary goals: (1) reducing traffic congestion through the addition of highway exchanges, bridges, and overpasses; (2) beautifying the city through the construction of monuments in the center of all major roundabouts; (3) renovating the city center and industrial zone and building 5,000 homes; and (4) making the city clean and healthy (ActuNiger 2017; République de Niger 2012). Who does this help? Most Niameyans do not own cars. State enterprises, including SPEN, helped to pay for monuments, gaining naming rights to roundabouts. As one of our interlocutors put it, reflecting the views of many others, "SPEN should spend their money to ensure that water flows through all standpipes even in the hot season instead of wasting it on monuments!" Nyala also involves demolishing *banco* houses visible from major boulevards (such as those in Pays Bas "too close" to their airport road and visible to international guests) and the destruction of 2,000 informal shops along major streets, as we witnessed in 2016–2017. From the state's point of view, this is part of renovation and making the city clean and healthy.

Conspicuous by their absence are any projects that might actually help the majority of the population, such as rebuilding the Petit Marché, building schools and staffing them with teachers who get paid reliably, building affordable health clinics, improving sanitation (such as by installing public toilets and starting garbage collection), and improving water access. In sum, Niamey Nyala seems to be primarily about modernizing and beautifying the city to please wealthy Niameyans and impress foreign visitors—it is all about appearances.

This vignette emphasizes several themes that have been woven into this book. Youssoufou and Zeinabou's struggles, the government's habit of ignoring the plight of the poor, and the misguided urban beautification projects of Niamey Nyala speak to the problems with governance and access we discussed in Chapter 1. Using Chatterjee's (2004: 38) terminology, Youssoufou and Zeinabou feel that they are merely part of the "population" of Niamey, and not full "rights-bearing citizens." Amazingly, water issues in any form are absent from Niamey Nyala. This story also illustrates the changing symbolic nature of water.

Water changed Youssoufou and Zeinabou's circumstances by providing them a means of earning money when Zeinabou's food table was no longer viable, but water was also changed *by* Youssoufou and Zeinabou's circumstances, as the destruction of the Petit Marché and subsequent diffusion of merchants into the surrounding neighborhood compromised water resources and eliminated Youssoufou's main customer base. For Youssoufou and Zeinabou and those involved in the water economies we included in this book, as well as thousands of other Niameyans, water is both life and profit. Changes in access to and governance of water have immediate and visible effects on their lives and livelihoods.

Youssoufou and Zeinabou's story is also an example of dualism in Nigérien society. Theirs is an example of Nigérien cultures of sharing, especially among the poor, juxtaposed with the greed, selfishness, and corruption that permeate the very government agencies that sit literally outside their compound. It speaks to ways water governance and access are connected to social realities and cultural symbolism. Addressing the problems in Zongo and the Petit Marché area would cost a very small fraction of Niger's national budget, yet despite the fact that government employees pass by these spaces daily on their way to and from work, problems persist. Corruption in Niger has local roots, but the problem with corruption, as Thomas Pogge (2008) argues, is that external forces can facilitate it. There is little incentive for government agents to fundamentally change the system if they can continue to benefit from a transnational system of inequality and are not directly affected by larger societal problems.

It is within this broader context of inequality and unequal access that Niamey's water regime operates. We have considered the various ways residents of Niamey access water and the hybrid economies and cultural symbolism that have emerged in light of historical and contemporary trends in water access, yet the ironies of these systems are not lost. A private connection to the piped network in Niamey, which provides the safest and healthiest form of drinking water in the city, is financially and systematically inaccessible to a large portion of the population, yet efforts to fill the gap in access require capital for upfront investment, whether it be in the form of pushcarts and containers or automated sachet machines.

Furthermore, inequalities in water access have opened opportunities for alternative incomes for Tuareg and Fulani migrants as *ga'ruwa*, women who own refrigerators that chill the sachets, children who sell sachets on the street, and the many men who make up the sachet water economy through production, delivery, and vending. Because of these complex, intricate systems of water access in Niamey, simply redistributing resources, water or any other resource, will not solve access problems. As Thomas Pogge (2008) recommends, we must change the world economic order that disadvantages the poor. Thousands of

Niamey residents support their families through the commoditization of water, which exists in part because state and privately controlled water systems are not equally distributed. New approaches to water governance must be mindful of this so that changes in water systems do not eliminate employment opportunities for those who currently depend on the system's inequitable distribution for their livelihoods.

At the very least, alternative means of employment should be considered as part of any project that significantly changes the alternative water-provision economies in Niamey and other Africa cities. In other words, the social and the political must be considered together. After all, compared to other parts of Niger and West Africa, water as a natural resource is not scarce in Niamey: the city sits on the Niger River, which has a year-round supply of surface freshwater. Rather, it is socioeconomics, social inequality, and governance that result in water access challenges and water scarcity for much of Niamey's population. Thus, those in power inflict structural violence on the "population" by manufacturing water scarcity in Niamey.

In his book *Drinking Water: A History*, James Salzman (2012: 10) calls drinking water "the story of empire" because "access to water is power." In Niamey, this empire is both local and global, domestic and international; it includes multinational corporations, regional decision makers, government agents, and private individuals. As Erik Swyngedouw (2004: 175) concludes, "Urban water is part and parcel of the political ecology of power that structures the functioning of the city." The existence of conditions like those in Zongo situated within line of sight of the people responsible for city and country governance is a visible example of purposeful abuse of power. Yet these conditions also speak to issues of social power, and while the world's poor are arguably the most vulnerable population with often the least amount of agency, the water economies we described in this book are examples of resilience and innovativeness that emerge in situations in which control over resources and economic systems becomes exclusionary.

Despite the inequitable distribution of water resources in Niamey, and the unequal pricing structures associated with access, there has been little grassroots organizing around water access and governance issues. This may, in part, be due to Niamey's water access issues compared to those in villages in Niger, where usually the only source of fresh water is subterranean, requiring wells and, more recently, boreholes to access. Because much of Niamey's population are migrants from rural Niger, the public standpipes and options for direct connections provide a more reliable water source, albeit an expensive one relative to village wells, from which water drawn is free. Subsequently, as of the time of our research, there were no water-related NGO projects in operation in Niamey. The NGOs with water-related agendas were almost exclusively working in rural areas, again perhaps because Niamey is seen as a place where water access involves fewer obstacles.

Furthermore, as we pointed out in Chapter 4, unlike other African cities, water is standardized within the hybridized vending system by vendors themselves. *Ga'ruwa* are informally organized and agree to use the same public standpipe so that their clients pay the same rate for water every day. The price of individual water sachets is the same throughout the city, regardless of who is selling it. This consistent pricing structure within Niamey perhaps communicates to NGOs and other groups that work in several countries (who are able to compare different water regimes) that the situation in Niamey, while not ideal, is better than in other places.

In general, however, water governance and access in Niamey is not pro-poor. The structural adjustment policies imposed by the World Bank and IMF came with conditions that prioritized privatization of, among other things, utilities in the name of efficiency and profitability. The existence of hybridized water economies and cultures like *ga'ruwa* vending and sachets and the cultural practices of water sharing reduce pressure on government and private entities to correct the systemic inequalities and invest in infrastructure. There is no relationship between urban sustainability and responsible water resource management in Niamey, certainly not one that takes cultural practices and materiality of water into account. The absence of water projects from Niamey Nyala provides a case in point. Water is life for all but profit for only some, and those profits are very unequally distributed.

Our explorations into water vending in this book reinforce water's connectivity, or all of the social realms influenced by water or where water is used, including political realms, infrastructural realms, governance realms, and material realms. Even climatic realms are relevant to our discussion, as climate change, which has shifted precipitation patterns and increased evaporation rates in the already vulnerable Sahel (and elsewhere), is anthropogenic. The totality of these connections helps create Niamey's "waterworlds" (see Orlove and Caton 2010; Hastrup and Hastrup 2016).

In these waterworlds, nature and infrastructure are inextricably linked. *Ga'ruwa* water delivery and sachet water are connected both to the city's utility networks and to the infrastructure, policies, and practices imbedded in the global economy. In a country like Niger, which is heavily dependent on natural resource extraction for income, global economic infrastructure provides most of the consumer goods and manufactured products that help make these local economies possible. The links between nature and infrastructure are fluid and changing, multiscalar, local and global, creating new symbolic meanings of water, life, and profit. Water is often contrasted with land. Land can be easily bounded and particular people can be excluded. Water is less easily bounded, as it moves and can be more easily shared (Orlove et al., 2016). Infrastructure and technology, however, have allowed us to bound water, especially that which is potable.

Subterranean water resources are particularly vulnerable to bounding because accessing these sources efficiently is heavily dependent on technology and infrastructure that is controlled by entities who wield the power and capital to employ such technology. Even water in Niamey is bounded, at least potable water. That which comes directly from the Niger River is contaminated and dangerous to drink, so most residents of Niamey must rely on infrastructure directly or indirectly to access potable water. Financial infrastructure and policies that govern prices of water extraction and expansion of networks further bound the safest drinking water in the city.

The relationship between nature and infrastructure speaks to broader issues of water security, or a threshold at which everyone has access to a sufficient quantity and quality of water for basic needs. Infrastructure like pipes allows more efficient access to water, but that water must be used sustainably, a paramount issue that calls upon responsible behavior by consumers as well as integrated water resource management (IWRM) entities. While approaches to water security vary significantly across disciplines (see Cook and Bakker 2012 for an overview), the issues most likely to affect Niamey in the future are availability and quantity of potable drinking water and, to a slightly lesser extent, of water for washing and bathing.

Impacts from the construction of the Kandadji Dam (in progress as we write this) have yet to be determined, but Niamey depends exclusively on the Niger River for its water supply. Construction of the Dam upstream from Niamey will reduce water flow, especially during the hot season. Large-scale irrigation projects proposed by the Millennium Development Corporation will require massive amounts of water from the Niger River, and most of these projects lie downstream from the dam and Niamey. Any impact on water flow implemented by neighboring Mali will have repercussions for Niger and Niamey as well.

In addition to water security, adaptive capacity is a desired outcome in IWRM (Lemos et al. 2016). Concerns over water security in Niamey, and efforts to improve security or reduce insecurity, must take into account the adaptive capacity of the city's population at all socioeconomic levels. In other words, any large-scale changes to governance in Niamey should consider the relationship between risk and development because these are key components for adaptive capacity (Lemos et al. 2016). While assessing adaptive capacities in Niamey is beyond the scope of this book, we mention it here in the conclusion because the ability to adapt to changes in water access has direct biological and economic outcomes—on life and profit. Changes in water security affect human capabilities, including good health (Nussbaum 2008), and thus a human capabilities approach to water security is needed, wherein the human right to water is of primary concern (Jepson et al. 2017).

Ensuring water security and expanding human capabilities are two conditions essential to good water governance. Those directly affected by changes in governance and access must be consulted and included in the process, and water conservation techniques must consider sociocultural knowledge and perceptions rather than just economic motivations (Bakker 2003; Loftus 2015). Niger, and other poor countries, desperately needs some form of democratic socialism wherein governments value democracy but also guarantee the poor's access to water, health care, food, and other key necessities. Water governance must be holistic in its implementation, taking into account water's commoditization, cultural symbolism, materiality, the global plastic capitalism within which water regimes operate, and water's connections to both local and global forces. Good water governance considers life over profit. Realities in Niamey, however, indicate that good water governance has yet to be achieved, although Niamey is not alone in this condition.

The water regime in Niamey, as we have described, is subject to top-down decision making, and the "top" is often external to Niger altogether (consider Veolia's involvement, for example). Problems arise when models that have achieved success in wealthy states are implemented in poor states where governments either do not or cannot distribute public services to the entire population. The ability to adequately extend public services is often compromised because poor states have inherited policies and programs from former colonial powers (see Chatterjee 2004 for more explanation). The line between "public" and "private" water is blurred in Niamey because, as Bakker (2010) describes, the Nigérien government simultaneously collaborates and competes with private enterprises and informal provision services, which could also be deemed "private" by some definitions. In short, the complex, overlapping, public-private water regime and access patterns in Niamey and the lack of attention to water's cultural symbolism exemplify Bakker's categories of state failure, market failure, and governance failure (see Chapter 1 for a discussion of these failures).

Access to safe drinking water is a fundamental human right, but someone must cover the cost of making water potable. Making Niger River water potable is expensive. States must develop policies and practices that make clean drinking water affordable for the poor. Global development objectives, like the Millennium Development Goals and the Sustainable Development Goals, are designed to move in that direction. However, successful fulfillment of these goals will require a global systematic reversal of the capitalist paradigms that have promoted the commoditization of water and distribution systems for decades.

Flint, Michigan is an important reminder that the water quality and access problems that often plague poor countries are inherent in the developed world as well. In a June 2018 news report from CBS (CBS 2018), Flint residents expressed a complete lack of trust in the very public entities that—in a country that has often considered itself an ideal democracy—were elected to protect and represent them. Even though the State of Michigan ended free bottled water distribution services in Flint in April 2018, residents continue to drink bottled water and put more trust in corporations (like Nestlé in Michigan) that benefit from and take advantage of this trust than they do in public entities and government officials.

Pro-poor water policies must prioritize people over profit, view human habitation holistically as waterworlds that include both natural and cultural forces, incorporate local knowledge about water and cultural elements in its use and distribution, and acknowledge water's materiality and links to the global trade in consumer goods, especially plastic. Public entities will need to work incredibly hard to regain the trust they have lost. Distributing clean water more equitably must be done in sustainable ways so that water resources providing this access remain available for future generations.

References

ActuNiger. 2017. "Infrastructures—Programme Niamey Nyala: Les Chantiers Qui On Changé le Visage de Niamey." 25 July. Accessed 15 June 2018 from https://www.actuniger.com/societe/13171-infrastructures-niamey-nyala-les-chantiers-qui-on-change-le-visage-de-Niamey.

Alidou, Ousseina. 2005. *Engaging Modernity: Muslim Women and the Politics of Agency in Postcolonial Niger*. Madison: University of Wisconsin Press.

Allan, Tony. 2003. "IWRM/IWRAM: A New Sanctioned Discourse?" Occasional Paper 50. SOAS Water Issues Group. School of Oriental and African Studies. King's College: University of London. Accessed 18 May 2018 from https://lwrg.files.wordpress.com/2014/12/iwram-a-new-sanctioned-discourse.pdf.

Annin, Peter. 2006. *The Great Lakes Water Wars*. Washington, DC: Island Press.

Appadurai, Arjun, ed. 1986. *The Social Life of Things: Commodities in Cultural Perspective*. Cambridge: Cambridge University Press.

Bakker, Karen. 2003. "Archipelagos and Networks: Urbanization and Water Privatization in the South." *The Geographical Journal* 169(4): 328–41.

_____. 2007. "Trickle Down? Private Sector Participation and the Pro-Poor Water Supply Debate in Jakarta, Indonesia." *Geoforum* 38: 855–68.

_____. 2010. *Privatizing Water: Governance Failure and the World's Urban Water Crisis*. Ithaca, NY: Cornell University Press.

Bakker, Karen, Michelle Kooy, Nur Endah Shofiani, and Ernst-Jan Martijn. 2008. "Governance Failure: Rethinking the Institutional Dimensions of Urban Water Supply to Poor Households." *World Development* 35(10): 1891–1915.

Bardasi, Elena and Quentin Wodon. 2008. "Who Pays the Most for Water? Alternative Providers and Service Cost in Niger." *Economics Bulletin* 9(20): 1–10.

Baron, Catherine. 2008. "Water Governance and Urban Local Development: An Analysis of Water Services Access in Sub-Saharan African Cities." In *Networks, Governance and Economic Development: Bridging Disciplinary Frontiers*, ed. Mari Jose Aranguren Querejeta, Cristina Iturrioz Landart, and James R. Wilson, 174–89. Cheltenham, UK: Edward Elgar.

_____. 2014. "Approvisionnement en Eau et Assainissement dans les Quartiers Défavorisés de Villes Africaines: État des Lieux Illustré à Travers les Cas de Ouagadougou et de Niamey." In *Eau Potable et Assainissement dans les Villes du Sud: La Difficile Intégration des Quartiers Défavorisés aux Politiques Urbaines*, ed. Frédéreic Naulet, Céline Gilquin, and Stéphanie Leyronas. Report #8 in the Collections Debats et Controverses. Agence Française de Développement. Accessed 22 May 2018 from https://www.pseau.org/outils/ouvrages/gre

t_eau_potable_et_assainissement_dans_les_villes_du_sud_la_difficile_inte gration_des_quartiers_defavorises_aux_politiques_urbaines_2014.pdf.
Baron, Catherine and Alain Bonnaissieux. 2011. "Les Enjeux de L'accès à L'eau en Afrique de l'Ouest: Diversité des Modes de Gouvernance et Conflits D'usages." *Mondes en Développement* 39(156): 17–32.
Baron, Catherine and Mahaman Tidjani Alou. 2011. "L'Accès à L'Eau en Afrique Subsaharienne: au-delà des Modèles, une Pluralité D'innovations Locales." *Mondes en Développement* 39(155): 7–22.
Bayliss, Kate and Rehema Tukai. 2011. "Services and Supply Chains: The Role of the Domestic Private Sector in Water Service Delivery in Tanzania." New York: United Nations Development Programme. Accessed 20 August 2015 from http://www.undp.org/content/dam/undp/library/Poverty%20Reduction/Inclusive%20developm ent/Tanzania-Water.pdf.
Berg, Sanford V. and Silver Mugisha. 2010. "Pro-Poor Water Service Strategies in Developing Countries: Promoting Justice in Uganda's Urban Project." *Water Policy* 12(4): 589–601. doi: 10.2166/wp.2010.120.
Berger, Arther Asa. 2014. *What Objects Mean: An Introduction to Material Culture*. 2nd ed. Walnut Creek, CA: Left Coast Press.
Bernus, Suzanne. 1969. *Particularismes Ethnique en Milieu Urbain: L'Example de Niamey*. Paris: Université de Paris, Institut d'Ethnologie, Musée de l'Homme.
Björkman, Lisa. 2015. *Pipe Politics, Contested Waters: Embedded Infrastructures of Millennial Mumbai*. Durham, NC: Duke University Press.
Bodley, John H. 2017. *Cultural Anthropology: Tribes, States, and The Global System*. Lanham, MD: Rowman & Littlefield.
Boelens, Rutgerd, Jaime Hoogesteger, Erik Swyngedouw, Jeroen Vos, and Pilippus Wester. 2016. "Hydrosocial Territories: A Political Ecology Perspective." *Water International* 41(1): 1–14.
Boko, Michel, Isabel Niang, Anthony Nyong, Coleen Vogel, Andrew Githeko, Mahmoud Medany, Balgis Osman-Elasha, Ramadjita Tabo, and Pius Yanda. 2007. "Africa." In *Climate Change 2007: Impacts, Adaptation and Vulnerability*. Contribution of Working Group 11 to the Fourth Assessment Report of the Intergovernmental Panel on Climate Change, ed. M. Parry, O. Canziani, J. Palutikof, P. van der Linden, and C. Hanson, 433–67. Cambridge: Cambridge University Press.
Bontianti, Abdou, Hilary Hungerford, Hassane H. Younsa, and Ali Noma. 2014. "Fluid Experiences: Comparing Local Adaptations to Water Inaccessibility in Two Disadvantaged Neighborhoods in Niamey, Niger." *Habitat International* 43: 283–92.
Bridge, Gavin and Adrian Smith. 2003. "Intimate Encounters: Economy–Culture–Commodity." *Environment and Planning D: Space and Society* 21: 257–68.
Brown, Oli, Anne Hammill, and Robert McLeman. 2007. "Climate Change as the 'New' Security Threat: Implication for Africa." *International Affairs* 83: 1141–54.
Buchli, Victor. 2002. "Introduction." In *The Material Culture Reader*, ed. Victor Buchli, 1–23. Oxford: Berg.

Cairncross, Sandy and Joanne Kinnear. 1991. "Water Vending in Urban Sudan." *Water Resources Development* 7(4): 267–73.
Callaway, Barbara J. 1984. "Ambiguous Consequences of the Socialisation and Seclusion of Hausa Women." *Journal of Modern African Studies* 22(3): 429–50.
Calvet, Louis-Jean. 1994. *Les Voix de la Ville: Introduction à la Sociolinguistique Urbaine*. Paris: Payot.
Caponera, Dante A. 2001. "La Propriété et le Transfert de L'Eau et des Terres dans l'Islam." In *La Gestion de L'Eau Selon l'Islam*, ed. N. I. Faruqui, A. K. Biswas, and M. J. Bino, 139–48. Paris: Karthala.
Carmody, Steve. 2018. "Bottled Water Distribution Ending in Flint." Michigan Public Radio. 6 April. Accessed 11 May 2018 from http://michiganradio.org/post/bottled-water-distribution-ending-flint.
CBS. 2018. "Flint Water Crisis: A Loss of Trust." 17 July. Accessed 20 June 2018 from https://www.cbsnews.com/news/the-flint-water-crisis-a-loss-of-trust/.
Chappal, Bill. 2018. "Michigan Oks Nestlé Water Extraction, Despite 80K+ Comments against It." National Public Radio. 3 April. Accessed 11 May 2018 from https://www.npr.org/sections/thetwo-way/2018/04/03/599207550/michigan-oks-nestl-water-extraction-despite-over-80k-public-comments-against-it.
Chatterjee, Partha. 2004. *Politics of the Governed: Reflections on Popular Politics in Most of the World*. New York: Columbia University Press.
Cohen, Michael A. 1989. "Urban Growth and Economic Development in the Sahel." World Bank: Staff Working Paper No. 315. Washington, DC: The World Bank.
Collins, H. and A. Wright. 2014. "Still Sparkling: The Phenomenon of Bottled Water—An Irish Context." *Journal of Marketing Management* 1: 15–31.
Commission for Africa. 2016. "Commissioners." Accessed 16 May 2018 from http://www.commissionforafrica.info/commissioners.
Conca, Ken. 2006. *Governing Water: Contentious Transnational Politics and Global Institution Building*. Cambridge, MA: MIT Press.
Cook, Christina and Karen Bakker. 2012. "Water Security: Debating an Emerging Paradigm." *Global Environmental Change* 22: 94–102.
Cook, Ian. 2004. "Follow the Thing: Papaya." *Antipode* 36(4): 642–64.
Coombe, Rosemary. 1996. "Embodied Trademarks: Mimesis and Alterity on American Cultural Frontiers." *Cultural Anthropology* 11(2): 202–25.
Dada, Ayokunle C. 2009. "Sachet Water Phenomenon in Nigeria: Assessment of the Potential Health Impacts." *African Journal of Microbiology Research* 3(1): 15–21.
Davey, Monica and Mitch Smith. 2016. "What Went Wrong in Flint?" *New York Times*. 3 March. Accessed 8 March 2018 from https://www.nytimes.com/interactive/2016/03/04/us/04flint-mistakes.html?rref=collection%2Fnewseventcollection%2Fflint-water-crisis&action=click&contentCollection=us&region=stream&module=stream_unit&version=latest&contentPlacement=9&pgtype=collection.
de Villiers, Marc and Sheila Hirtle. 2002. *Sahara: A Natural History*. New York: Walker.

Dearden, Nick. 2012. "Greece Can Do without the 'Sympathy' the IMF Has Shown Niger." *The Guardian*. Accessed 29 June 2012 from https://www.theguardian.com/commentisfree/2012/may/29/greece-sympathy-imf-niger.

Decalo, Samuel. 1989. *Historical Dictionary of Niger*. 2nd ed. London: Scarecrow Press.

Demographia. 2015. "Demographia World Urban Areas, 11th Annual Edition: 2015:01." Accessed 1 February 2016 from http://www.urbangateway.org/system/files/documents/urbangateway/db-worldua.pdf.

Diallo, Souleymane. 2016. "'We must all go to the Hangar': Performing Bellah Group Membership in the Refugee Camp in Abala, Niger." *Journal for Islamic Studies* 35: 43–69.

DiFrancesco, Darryn Anne and Nathan Young. 2010. "Seeing Climate Change: The Visual Construction of Global Warming in Canadian National Print Media." *Cultural Geographies* 18(4): 517–36.

Dill, Brian and Ben Crow. 2014. "The Colonial Roots of Inequality: Access to Water in Urban East Africa." *Water International* 39(2): 187–200. doi: 10.1080/02508060.2016.1134898.

Douglas, Mary. 1970. *Natural Symbols: Explorations in Cosmology*. New York: Pantheon Books.

Drewal, Henry, ed. 2008. *Sacred Waters: Arts of Mami Wata and Other Divinities in Africa and the Diaspora*. Bloomington: Indiana University Press.

Dupire, Marguerite. 1962. *Peuls Nomades: Étude Descriptive des WoDaaBe du Sahel Nigérien*. Paris: Institut d'Ethnologie.

Faruqui, Naser I., Asit K. Biswas, and Murad J. Bino, eds. 2001. *Water Management in Islam*. Tokyo: United Nations University Press.

Fisher, Michael B., Ashley R. Williams, Mohammad F. Jalloh, George Saquee, Robert. E. S. Bain, and Jamie K. Bartram. 2015. "Microbiological and Chemical Quality of Packaged Sachet Water and Household Stored Drinking Water in Freetown, Sierra Leone," *PLoS ONE* 10(7): e0131772.

Fonger, Ron. 2018. "Flint Mayor Wants State-Funded Bottled Water until All Lead Lines Are Gone." MLive. 5 March. Accessed 8 March 2018 from http://www.mlive.com/news/flint/index.ssf/2018/03/flint_mayor_wants_state-funded.html.

Gado, Boureima Alpha. 1997. *Niamey: Garin Kaptan Salma (Histoire d'une Ville)*. Niamey, Niger: Nouvelle Imprimerie du Niger.

Geertz, Clifford. 1973. *The Interpretation of Cultures*. New York: Basic Books.

Gibbon, Peter and Stefano Ponte. 2005. *Trading Down: Africa, Value Chains, and the Global Economy*. Philadelphia: Temple University Press.

Giuffrida, Alessandra. 2010. "Tuareg Networks: An Integrated Approach to Mobility and Stasis." In *Tuareg Society within a Globalized World: Saharan Life in Transition*, ed. Anja Fischer and Ines Kohl, 23–40. London: Tauris Academic Studies.

Gleick, Peter. 2003. "Global Freshwater Resources: Soft-Path Solutions for the 21st Century." *Science* 302(28 Nov.): 1524–28.

———. 2010. *Bottled and Sold: The Story behind Our Obsession with Bottled Water*. Washington, DC: Island Press.

Glenzain, Jessica. 2017. "Nestle Pays $200 a Year to Bottle Water near Flint—Where Water Is Undrinkable." *The Guardian*. 29 September. Accessed 11 May 2018 from https://www.theguardian.com/us-news/2017/sep/29/nestle-pays-200-a-year-to-bottle-water-near-flint-where-water-is-undrinkable.

Global Water Partnership. 2018. "What Is IWRM?" Accessed 14 May 2018 from https://www.gwp.org/en/GWP-CEE/about/why/what-Is-iwrm/.

Goldman, Michael. 2007. "How 'Water for All!' Policy Became Hegemonic: The Power of the World Bank and Its Transnational Policy Networks." *Geoforum* 38(5): 786–800.

Graham, Stephen and Simon Marvin. 2001. *Splintering Urbanism: Networked Infrastructures, Technological Mobilities, and the Urban Condition*. New York: Routledge.

Grant, Richard. 2015. *Africa: Geographies of Change*. New York: Oxford University Press.

Guardian, The. 2015. "Where Is the Fastest Growing City in the World?" Accessed 15 May 2018 from http://www.theguardian.com.

Gupta, Sanjay, Ben Tinker, and Tim Hume. 2016. "'Our mouths were ajar': Doctor's Fight to Expose Flint's Water Crisis." CNN. 22 January. Accessed 8 March 2018 from https://www.cnn.com/2016/01/21/health/flint-water-mona-hanna-attish/.

Hall, Bruce. 2011. *A History of Race in Muslim West Africa, 1600–1960*. Cambridge: Cambridge University Press.

Hall, David and Emanuele Lobina. 2007. "Profitability and the Poor: Corporate Strategies, Innovation and Sustainability." *Geoforum* 38(5): 772–85.

Hallam, W. K. R. 1966. "The Bayajida Legend in Hausa Folklore." *The Journal of African History* 7(1): 47–60.

Hamidou, Issaka Maga and Sani Ali. 2005. *Impacts Sectoriels de la Croissance Démographique dans un Contexte de Strategie de Réduction de la Pauverté: Perspectives Dérivées 2005–2050*. Niamey, Niger: République du Niger, Ministère de la Population et de l'Action Sociale.

Hanchett, Suzanne, Tofazzel Hossain Monnju, Kazi Rozana Akhter, Shireen Akhter, and Anwar Islam. 2014. *Water Culture in South Asia: Bangladesh Perspectives*. Pasadena, CA: Development Resources Press.

Hansen, Karen Tranberg in collaboration with Anne Line Dalsgaard, Katherine V. Gough, Ulla Ambrosius Madsen, Karen Valentin, and Norbert Wildermuth. 2008. *Youth and the City in the Global South*. Bloomington: Indiana University Press.

Harding, Karol. 1996. "He Is Coming, She Is Coming Guerda, The Tuareg Blessing Dance." *The Best of Habibi* 15(3): 1–8.

Hastrup, Kirsten and Frida Hastrup. 2016. "Introduction: Waterworlds at Large." In *Waterworlds: Anthropology in Fluid Environments*, ed. Kirsten Hastrup and Frida Hastrup, 1–22. New York: Berghahn Books.

Hawkins, Gail, Emily Potter, and Kane Race. 2015. *Plastic Water: The Social and Material Life of Bottled Water*. Cambridge, MA: MIT Press.

Hill, Polly. 1969. "Hidden Trade in Hausaland." *Man* 4(3): 392–409.

Humphrey, John and Hubert Schmitz. 2002. "Developing Country Firms in the World Economy: Governance and Upgrading in Global Value Chains," INEF

Report 61. Institut für Entwicklung und Frieden der Gerhard-Mercator-Universität Duisburg für Entwicklung und Frieden. Accessed 16 October 2017 from http://citeseerx.ist.psu.edu/viewdoc/download?doi=10.1.1.557.1063&rep=rep1&type=pdf.

Hungerford, Hilary. 2012. "Water, Cities, Bodies: A Relational Understanding of Niamey, Niger." PhD diss. University of Kansas.

Hungerford, Hilary and Sarah L. Smiley. 2016. "Comparing Colonial Water Provision in British and French Africa." *Journal of Historical Geography* 52: 74–83.

Ibrahim, Mustapha, Musa Umaru, and Akindele Akinsoji. 2015. "Qualitative Assessment of Sachet and Bottled Water Marketed in Bauchi Metropolis, Nigeria." *Journal of Food Process Engineering* 37: 11–23.

Ibrahim, Solava. 2006. "From Individual to Collective Capabilities: The Capability Approach as a Conceptual Framework for Self-help." *Journal of Human Development* 7(3): 397–416.

Idrissa, Abdourahmane. 2009. "The Invention of Order: Republican Codes and Islamic Law in Niger." PhD diss. University of Florida.

INS (Institut National de la Statistique). 2017. "Chiffres: Éditions Annuelle." Accessed 27 May 2018 from http://www.stat-niger.org.

Institute for Health Metrics and Evaluation (IHME). 2018. "Niger." Accessed 1 June 2018 from http://www.healthdata.org/niger.

Intergovernmental Panel on Climate Change Working Group II. 2014. "Climate Change 2014: Impacts, Adaptations, and Vulnerability." Accessed 19 June 2018 from http://www.ipcc.ch/publications_and_data/publications_and_data_reports.shtml.

Jepson, Wendy, Jessica Budds, Laura Eichelberger, Leila Harris, Emma Norman, Kathleen O'Reilly, Amber Pearson, Sameer Shah, Jamie Shinn, Chad Staddon, Justin Stoler, Amber Wutich, and Sera Young. 2017. "Advancing Human Capabilities for Water Security: A Relational Approach." *Water Security* 1: 46–52.

Joffe, Hélèn. 2008. "The Power of Visual Material: Persuasion, Emotion and Identification." *Diogenes* 55(1): 84–93.

Kaplinsky, Raphael. 2000. "Globalisation and Unequalisation: What Can Be Learned from Value Chain Analysis?" *Journal of Development Studies* 37(2): 117–46.

Kaplinsky, Raphael and Mike Morris. 2001. "A Handbook for Value Chain Research," Ottawa: International Development Research Centre (IDRC). Accessed 26 July 2016 from https://www.ids.ac.uk/ids/global/pdfs/ValuechainHBRKMMNov2001.pdf.

Kariuki, Mukami, Rosemary Rop, and Madori Makino. 2014. "Do Pro-Poor Policies Increase Water Coverage?" The Water Blog by the World Bank. Accessed 24 May 2018 from https://blogs.worldbank.org/water/do-pro-poor-policies-increase-water-coverage.

Keenan, Jeremy. 2013. *The Dying Sahara: US Imperialism and Terror in Africa*. London: Pluto Press.

Keener, Sarah, Manuel Luengo, and Sudeshna Banerjee. 2010. "Provision of Water to the Poor in Africa: Experience with Water Standposts and the

Informal Water Sector." Policy Research Working Paper 5387. The World Bank (Africa Region): Sustainable Development Division (July 2010). Accessed 24 May 2018 from http://documents.worldbank.org/curated/en/421921468191675047/pdf/wps5387.pdf.

Keough, Sara Beth and Scott M. Youngstedt. 2014. "The Material Culture of Water: Transportation, Storage, and Consumption in Niamey, Niger." *Focus on Geography* 57(4): 152–63.

———. 2018. "'Pure Water' in Niamey, Niger: The Backstory of Sachet Water in a Landscape of Waste." *Africa: Journal of the International Africa Institute* 88(1): 38–62.

Kjellén, Marianne. 2000. "Complimentary Water Systems in Dar es Salaam, Tanzania: The Case of Water Vending." *Water Resources Development* 16(1): 143–54.

———. 2006. "From Public Pipes to Private Hands: Water Access and Distribution in Dar es Salaam, Tanzania." PhD diss. Stockholm University.

Kooy, Michelle and Karen Bakker. 2014. "Post-Colonial Pipes: Urban Water Supply in Colonial and Contemporary Jakarta." In *Cards, Conduits, and Kampongs: The Modernization of the Indonesian City, 1920–1960*, ed. F. Colombijin and J Coté, 63–86. Leiden: Brill.

Laurie, Nina. 2007. "Introduction: How to Dialogue for Pro-Poor Water." *Geoforum* 38(5): 753–55.

Lecocq, Baz. 2011. *Disputed Desert: Decolonisation, Competing Nationalisms and Tuareg Rebellions in Northern Mali*. Leiden: Brill.

Lee, Margaret. 2014. *Africa's World Trade: Informal Economies and Globalization from Below*. London: Zed Books.

Lemos, Maria Carmen, David Manuel-Navarrete, Bram Leo Willems, Rolando Diaz Caravantes, and Robert G. Varady. 2016. "Advancing Metrics: Models for Understanding Adaptive Capacity and Water Security." *Current Opinion in Environmental Sustainability* 21: 52–57.

Lévi-Strauss, Claude. 1966. *The Savage Mind*. Chicago: University of Chicago Press.

Limbert, Mandana E. 2010. *In the Time of Oil: Piety, Memory, and Social Life in an Omani Town*. Stanford: Stanford University Press.

Lindell, Ilda, ed. 2010. *Africa's Informal Workers: Collective Agency, Alliances and Transnational Organizing in Africa*. London: Zed Books.

Linton, Jamie and Jessica Budds. 2014. "The Hydro-Social Cycle: Defining and Mobilizing a Relational-Dialectical Approach to Water." *Geoforum* 57: 170–80.

Loftsdóttir, Kristín. 2008. *The Bush Is Sweet: Identity, Power and Development among WoDaaBe Fulani in Niger*. Uppsala, Sweden: Nordiska Affrkainstitutet.

Loftus, Alex. 2015. "Water (In)Security: Securing the Right to Water." *The Geographical Journal* 181(4): 350–56.

Lury, Celia. 2004. *Brands: The Logos of the Global Economy*. New York: Routledge.

MacDonald, A. M., H. C. Bonsor, B. É. Ó. Dochartaigh, and R. G. Taylor. 2012. "Quantitative Maps of Groundwater Resources in Africa." *Environmental Research Letters* 7(2): 1–7.

Macdonald, Kate. 2007. "Globalising Justice within Coffee Supply Chains? Fair Trade, Starbucks and the Transformation of Supply Chain Governance." *Third World Quarterly* 28(4): 793–812.
Maiga, Taibou Adamou. 2016. "Delivering Water and Sanitation Services in Niger: Challenges and Results." World Bank Water Blog. Accessed 1 May 2018 from http://blogs.worldbank.org/water/delivering-water-and-sanitation-services-niger-challenges-and-results.
Martinez, Michael. 2016. "Flint, Michigan: Did Race and Poverty Factor into the Water Crisis?" CNN. Accessed 8 March 2018 from https://www.cnn.com/2016/01/26/us/flint-michigan-water-crisis-race-poverty/index.html.
Masquelier, Adeline. 2008. "Water Spirits in Water-less Places." In *Sacred Waters: Arts of Mami Wata and Other Divinities in Africa and the Diaspora*, ed. Henry John Drewal, 75–86. Bloomington: Indiana University Press.
Mauss, Marcel. 1950. (republished 1990). *The Gift: The Form and Reason for Exchange in Archaic Societies*. London: Routledge.
Meagher, Kate. 2010. *Identity Economics: Social Networks and the Informal Economy in Nigeria*. Rochester, NY: James Currey.
Miller, Daniel. 2005. "Materiality: An Introduction." In *Materiality*, ed. Daniel Miller, 1–50. Durham, NC: Duke University Press.
Ministerial Declaration. 2000. 2nd World Water Forum. Accessed 8 May 2018 from http://www.waternunc.com/gb/secwwf12.htm.
Modern Ghana. 2011. "Drought Threatens Niger." Accessed 22 May 2018 from https://www.modernghana.com/news/344002/drought-threatens-niger.html.
Motcho, M. Kokou Henri. 1992. "Problèmes de Croissance Urbaine dans la Communauté Urbaine de Niamey." Niamey, Niger: Colloque de Niamey: "L'Environment Urbain," Association Internationale des Maires et Responsables des Capitales et Métropoles Partièllement ou Entièrement Francophones.
Mtisi, Sobona and Alan Nicol. 2015. "Water Politics in Eastern and Southern Africa." In *Water and Development: Good Governance after Neoliberalism*, ed. Ronaldo Munck, Narathius Asingwire, Honor Fagan, and Consolata Kabonesa, 84–103. London: Zed Books.
Munck, Ronaldo. 2015. "Water, Development and Good Governance." In *Water and Development: Good Governance after Neoliberalism*, ed. Ronaldo Munck, Narathius Asinguire, Honor Fagan, and Consolata Kabonesa, 11–29. London: Zed Books.
Myers, Garth. 2011. *African Cities: Alternative Visions of Urban Theory and Practice*. New York: Zed Books.
Ngmekpele Cheabu, Benjamin S. and James Hawkins Ephraim. 2014. "Consumers' Perception of Quality and Health Benefits of Sachet Drinking Water: Evidence from Obuasi in the Ashanti Region of Ghana." *Developing Country Studies* 4(17): 66–77.
Niasse, Madiodio. 2005. "Climate-Induced Water Conflict Risks in West Africa: Recognizing and Coping with Increasing Climate Impacts on

Shared Watercourses." Paper presented at Human Security and Climate Change: An International Workshop, Oslo, Norway, 21–23 June 2005. Accessed 9 May 2018 from https://www.researchgate.net/profile/Madiodio_Niasse/publication/237699436_Climate-Induced_Water_Conflict_Risks_in_West_Africa_Recognizing_and_Coping_with_Increasing_Climate_Impacts_on_Shared_Watercourses/links/5440fe550cf2a76a3cc60e7c/Climate-Induced-Water-Conflict-Risks-in-West-Africa-Recognizing-and-Coping-with-Increasing-Climate-Impacts-on-Shared-Watercourses.pdf.

Nickson, Andrew and Richard Franceys. 2003. *Tapping the Market: The Challenge of Institutional Reform in the Urban Water Sector*. Gordonsville, VA: Palgrave.

Njoh, Ambe J. 2008. "Colonial Philosophies, Urban Space, and Racial Segregation in British and French Colonial Africa." *Journal of Black Studies* 38(4): 579–99.

Njoh, Ambe J. and Fenda A. Akiwumi. 2011. "The Impact of Colonization on Access to Improved Water and Sanitation Facilities in African Cities." *Cities* 28: 452–60.

Nussbaum, Martha. 2008. "Human Capabilities, Female Human Beings." In *Global Ethics: Seminal Essays Volume II—Global Responsibilites*, ed. Thomas Pogge and Keith Horton, 495–551. St. Paul, MN: Paragon House.

Nwadike, Peter. 2012. "How Pure Is Our 'Pure Water?'" Ezine articles from http://ezinearticles.com/?How-Pure-Is-Our-Pure-Water?&id=7382860. Accessed 16 October 2017 from http://newsciencejournalism.com/11/2012/how-pure-is-nigerias-pure-water.

Olivier de Sardan, Jean-Pierre. 1982. *Concepts et Conceptions Songhay-Zarma*. Paris: Nubia.

Oosting, Jonathan and Beth LeBlanc. 2018. "Snyder Ending Bottled Water for Flint." Detroit News. 6 April. Accessed 11 May 2018 from https://www.detroitnews.com/story/news/politics/2018/04/06/snyder-ends-bottled-water-flint/33612873/.

Opryszko, Melissa, Haiou Huang, Kurt Soderland, and Kellogg Schwabb. 2009. "Data Gaps in Evidence-Based Research on Small Water Enterprises in Developing Countries." *Journal of Water and Health* 7(4): 609–22.

Orlove, Ben and Steven C. Caton. 2010. "Water Sustainability: Anthropological Approaches and Prospects." *Annual Review of Anthropology* 39: 401–15.

Orlove, Ben, Carla Roncoli, and Brian Dowd-Uribe. 2016. "Fluid Entitlements: Constructing and Contesting Water Allocations in Burkina Faso, West Africa." In *Waterworlds: Anthropology in Fluid Environments*, ed. Kirsten Hastrup and Frida Hastrup, 46–74. New York: Berghahn Books.

Ortner, Sherry B. 1973. "On Key Symbols." *American Anthropologist* 75(5): 1338–46.

Osumanu, Issaka Kanton. 2008. "Private Sector Participation in Urban Water and Sanitation Provision in Ghana: Experiences from the Tamale Metropolitan Area (TMA)." *Environmental Management* 42: 102–10.

Osumanu, Issaka Kanton and Lukeman Abdul-Rahim. 2008. "Enhancing Community-Driven Initiatives in Urban Water Supply in Ghana." Proceedings of the 33rd WEDC International Conference on Access to Sanitation and Safe

Water: Global Partnerships and Local Actions, Accra, Ghana. Loughborough, UK: Loughborough University of Technology.

Page, Ben. 2005. "Paying for Water and the Geography of Commodities." *Transactions of the Institute for British Geographers* 30(3): 293–306.

Parnell, S. and R. Walawege. 2011. "Africa South of the Saharan Urbanization and Global Environmental Change." *Global Environmental Change* 21: 12–20.

Patel, Prachi. 2018. "Stemming the Plastic Tide: 10 Rivers Contribute Most of the Plastic in the Oceans." *Scientific American*. 1 February. Access 28 June 2018 from https://www.scientificamerican.com/article/stemming-the-plastic-tide-10-rivers-contribute-most-of-the-plastic-in-the-oceans/.

Penny, Joe. 2018. "Drones in the Sahara: A Massive U.S. Drone Base Could Destabilize Niger—And May Even Be Illegal under Its Constitution." 18 February. *The Intercept*. Accessed 7 June 2018 from https://theintercept.com/2018/02/18/niger-air-base-201-africom-drones/.

Pieterse, Edgar and Susan Parnell. 2014. "Africa's Urban Revolution in Context." In *Africa's Urban Revolution*, ed. Susan Parnell and Edgar Pieterse, 1–17. New York: Zed Books.

Pogge, Thomas. 2008. *World Poverty and Human Rights: Cosmopolitan Responsibilities and Reforms*. 2nd ed. Cambridge, UK: Polity Press.

Probyn, Justin. 2016. "The Indian Entrepreneur Who Started a Bottled Water Company in Niger." How We Made It in Africa. 7 November. Accessed 14 June 2018 from https://www.howwemadeitinafrica.com/indian-entrepreneur-started-bottled-water-company-niger/56598/.

Rasmussen, Susan. 1992. "Ritual Specialists, Ambiguity and Power in Tuareg Society." *Man* 27(1): 105–28.

_____. 1998. "Ritual Powers and Social Tensions as Moral Discourse among the Tuareg." *American Anthropologist* 100(2): 458–68.

_____. 2009. "Pastoral Nomadism and Gender: Status and Prestige, Economic Contribution, and Division of Labor among the Tuareg of Niger." In *Gender in Cross-Cultural Perspective*, 5th ed., ed. Caroline B. Brettell and Carolyn F. Sargent, 280–93. Upper Saddle River, NJ: Prentice Hall.

Regis, Helen A. 2003. *Fulbe Voices*. Boulder, CO: Westview.

ReliefWeb. 2006. "Niger: Praying for Rain." 28 June. Report from IRIN. Accessed 22 May 2018 from https://reliefweb.int/report/niger/niger-praying-rain.

République du Niger. 2010. *La Population du Niger en 2010*. Niamey, Niger: L'Institut National de la Statitistique, Ministère de l'Économie et des Finances.

_____. 2012. "Niamey Nyala." Niamey, Niger: Ministère de l'Urbanisme du Logement et de l'Assainissemant.

Ricketts, Katie D., Calum G. Turvey, and Miguel I. Gómez. 2014. "Value Chain Approaches to Development: Smallholder Farmer Perceptions of Risk and Benefits across Three Cocoa Chains in Ghana." *Journal of Agribusiness in Developing and Emerging Economies* 4(1): 2–22.

Rizzo, Matteo. 2013. "Informalisation and the End of Trade Unionism as We Knew It? Dissenting Remarks from a Tanzanian Case Study." *Review of African Political Economy* 40(136): 290–308.

Robeyns, Ingrid. 2005. "The Capability Approach: A Theoretical Survey." *Journal of Human Development* 6(1): 93–114.

Salzman, James. 2012. *Drinking Water: A History*. New York: Overlook Press.

Sambu, Daniel. 2016. "Impact of Global Initiatives on Drinking Water Access in Africa." *African Geographical Review* 35(2): 151–67.

Sassen, Saskia. 2007. *Deciphering the Global: Its Scales, Spaces and Subjects*. London: Routledge.

Schildkrout, Enid. 1982. "Dependency and Autonomy: The Economic Activities of Secluded Hausa Women in Kano." In *Women and Work in Africa*, ed. E. G. Bay, 55–83. Boulder, CO: Westview Press.

Schwartz, Klaas, Mireia Tutusaus Luque, Maria Rusca, and Rhodante Ahlers. 2015. "(In)formality: The Meshwork of Water Service Provisioning." *WIREs Water* 2: 31–36.

Sen, Amartya. 1999. *The Capability Approach: A Theoretical Survey*. New York: Knopf.

Shack, William A. and Elliott P. Skinner, eds. 1979. *Strangers in African Societies*. Berkeley: University of California Press.

Sidikou, Hamidou A. 1980. "Niamey, Étude de Geographie Socio-urbaine." Doctoral diss. Université de Rouen Haute-Normandie.

Simone, AbdouMaliq. 2004. "People as Infrastructure: Intersecting Fragments in Johannesburg." *Public Culture* 16(3): 407–29.

Slaymaker, Tom and Robert Bain. 2017. "Access to Drinking Water around the World—in Five Infographics." *The Guardian*. Accessed 17 March from https://www.theguardian.com/global-development-professionals-network/2017/mar/17/access-to-drinking-water-world-six-infographics.

Smiley, Sarah L. 2013. "Complexity of Water Access in Dar es Salaam, Tanzania." *Applied Geography* 41: 132–38.

Smiley, Sarah L. and Francis T. Koti. 2010. "Introduction to Special Issue: Africa's Spaces of Exclusion." *Africa Today* 56(3): v–ix.

Stoler, Justin. 2017. "From Curiosity to Commodity: A Review of the Evolution of Sachet Drinking Water in West Africa." *WIREs Water* 4(3): e1206. doi: 10.1002/wat2.1206.

Stoler, Justin, Raymond A. Tutu, Hawa Ahmed, Lady A. Frimpong, and Mohammad Bello. 2014. "Sachet Water Quality and Brand Reputation in Two Low-Income Urban Communities in Greater Accra, Ghana." *American Journal of Tropical Medicine* 90(2): 272–78.

Stoler, Justin, John Weeks, and Gunther Fink. 2012. "Sachet Drinking Water in Ghana's Accra-Tema Metropolitan Area: Past, Present, and Future." *Journal of Water Sanitation Hygiene Development* 2(4): 1–24.

Stoller, Paul. 2002. *Money Has No Smell: The Africanization of New York City*. Chicago, IL: University of Chicago Press.

Stoller, Paul and Cheryl Olkes. 1987. *In Sorcery's Shadow: A Memoir of Apprenticeship among the Songhay of Niger*. Chicago: University of Chicago Press.

Strang, Veronica. 2004. *The Meaning of Water*. Oxford: Berg.

_____. 2013. *Gardening the World: Agency, Identity and the Ownership of Water*. New York: Berghahn Books.

———. 2015. *Water: Nature and Culture*. London: Reaktion Books.
Swatuk, Larry A. 2015. "Can IWRM Float on a Sea of Underdevelopment? Reflections on Twenty-Plus Years of 'Reform' in Sub-Saharan Africa." In *Water and Development: Good Governance after Neoliberalism*, ed. Ronaldo Munck, Narathius Asingwire, Honor Fagan, and Consolata Kabonesa, 60–83. London: Zed Books.
Swyngedouw, Erik. 2004. *Social Power and the Urbanization of Water: Flows of Power*. Oxford, UK: Oxford University Press.
Tidjani Alou, Mahaman. 2005. "Le Partenariat Public-Privé dans le Secteur de l'Eau au Niger: Autopsie d'une Réforme." *Annuaire Suisse de Politique de Développement* 24(2): 161–77.
Tsing, Anna. 2013. "Sorting Out Commodities: How Capitalist Value Is Made through Gifts." *HAU: Journal of Ethnographic Theory* 3(1): 21–43.
UN (United Nations: Population Division of the Department of Economic and Social Affairs). 2012. "World Urbanization Prospects: Niamey." Accessed 28 June 2015 from http://knoema.com/UNWUP2011R/world-urbanization-prospects-the-2011-revision-urban-agglomerations?tsId=1002200.
UNDP (UN Development Programme). "Sustainable Development Goals." 2016a. Accessed from http://www.undp.org/content/undp/en/home/sustainable-development-goals.html.
———. "Human Development Index." 2016b. Niger. Accessed 31 March 2019 from http://hdr.undp.org/en/countries/profiles/NER.
United Nations Human Development Reports. 2016. "Niger Country Profile." Accessed 10 June 2018 from http://hdr.undp.org/en/countries/profiles/NER.
US Bureau of the Census. 2016. "Quick Facts: Flint City, Michigan." Accessed 31 May 2018 from https://www.census.gov/quickfacts/fact/table/flintcitymichigan/PST045216.
USAID (US Agency for International Development). 2010. "Niger Water and Sanitation Profile." Accessed 20 August 2015 from http://www.washplus.org/sites/default/files/niger.pdf.
———. 2010. "Niger: Water and Sanitation Profile." Accessed 7 May 2018 from http://www.washplus.org/sites/default/files/niger.pdf.
———. 2018. "Power Africa Fact Sheet—Niger." Accessed 7 June 2018 from https://www.usaid.gov/powerafrica/niger.
Varisco, Daniel Martin. 1983. "Sayl and Ghayl: The Ecology of Water Allocation in Yemen." *Human Ecology* 11(4): 365–83.
Veolia, Inc. 2017. "Société d'Exploitation des Eaux du Niger (SEEN)—Niger" Accessed 8 May 2018 from https://www.veolia.com/africa/en/our-clients/societe-dexploitation-des-eaux-du-niger-seen-niger.
Wade, Jeffrey. 2012. "The Future of Urban Water Services in Latin America." *Bulletin of Latin American Research* 31(2): 207–21.
Wagner, John R. 2013. "Introduction." In *The Social Life of Water*, ed. John R. Wagner, 1–14. New York: Berghahn Books.
———, ed. 2013. *The Social Life of Water*. New York: Berghahn Books.
Walentowitz, Saskia. 2011. "Women of Great Weight: Fatness, Reproduction, and Gender Dynamics in Tuareg Society." In *Fatness and the Maternal*

Body: Women's Experiences of Corporeality and the Shaping of Social Policy, ed. Maya Unnithan-Kumar and Soraya Tremayne, 71–97. New York: Berghahn Books.

Wall, L. Lewis. 1988. *Hausa Medicine: Illness and Well-Being in a West African Culture*. Durham, NC: Duke University Press.

Wescoat, James L. Jr., Lisa Headington, and Rebecca Theobald. 2007. "Water and Poverty in the United States." *Geoforum* 38(5): 801–14.

Whittington, Dale, Donal Lauria, Daniel A. Okun, and Xinming Mu. 1989. "Water Vending Activities in Developing Countries: A Case Study of Ukunda, Kenya." *Water Resources Development* 5(3): 158–68.

WHO (World Health Organization). 2012. "Millennium Development Goal Drinking Water Target Met, Sanitation Target Still Lagging far Behind." WHO Media Center. Accessed 5 March 2012 from http://www.who.int/mediacentre/news/releases/2012/drinking_water_20120306/en/.

———. 2015. "Niger: WHO Statistical Profile." Accessed 7 May 2018 from http://www.who.int/gho/countries/ner.pdf?ua=1.

Wilk, Richard. 2006. "Bottled Water, the Pure Commodity in the Age of Branding." *Journal of Consumer Culture* 6(3): 303–25.

Williams, Hugh. 2015. "COP 21: Five Ways Climate Change Could Affect Africa." BBC News Africa. 11 December. Accessed 9 May 2018 from http://www.bbc.com/news/world-africa-35054300.

Wilson Fall, Wendy. 2015. "Local Texts, Rumor and Ethnic Ideologies: The Amazigh Community and Its Border Identities." In *Sahara Crossroads: View from the Desert Edge*, ed. Jennifer Yanco, 68–85. Oran, Algeria: Centre de Recherche en Anthropologie Sociale (CRASC), University of Oran.

Wittfogel, Karl A. 1957. *Oriental Despotism: A Comparative Study of Total Power*. New Haven, CT: Yale University Press (French Translation: *Le Despotisme Oriental. Étude Comparative du Pouvoir Total*. Paris: Éditions de minuit, 1964).

World Bank. 2017. "Niger Country Overview." 5 December. Accessed 7 May 2018 from http://www.worldbank.org/en/country/niger/overview.

World Water Forum. 2000. Accessed 8 May 2018 from http://www.worldwaterforum5.org/index.php?id=1961&L=1%2Findex.php%3Fcibl%20target%3D%20title%3D%20target%3D.

WSP (Water and Sanitation Program). 2011. "Water Supply and Sanitation in Niger." Accessed 15 January 2018 from https://www.wsp.org/sites/wsp/files/publications/CSO-Niger.pdf.

Yanco, Jennifer J. 2014. *Misremembering Dr. King: Revisiting the Legacy of Martin Luther King Jr*. Bloomington: Indiana University Press.

Youngstedt, Scott M. 2013. *Surviving with Dignity: Hausa Communities of Niamey, Niger*. Lanham, MD: Lexington Books.

Youngstedt, Scott M., Sara Beth Keough, and Cheiffou Idrissa. 2016. "Water Vendors in Niamey: Considering the Economic and Symbolic Nature of Water." *African Studies Quarterly* 16(2): 27–46.

Zdanowicz, Christine. 2016. "Flint Family Uses 151 Bottles of Water per Day." CNN. 16 March. Accessed 11 May 2018 from https://www.cnn.com/2016/03/05/us/flint-family-number-daily-bottles-of-water/index.html.

Zifiti. 2018. "Zamzam 500 ml Bottle Water from Mecca Makkah Saudi Arabia Zamzam Free Shipping." Accessed 20 May 2018 from https://zifiti.com.

Zylberberg, Ezequiel. 2013. "Bloom or Bust? A Global Value Chain Approach to Smallholder Flower Production in Kenya." *Journal of Agribusiness in Developing and Emerging Economies* 3(1): 4–26.

Index

adaptive capacity 157–58
archipelagos 26, 62, 101

Bayajida 34
block tariffs 15, 29
Bodley, John 9
Boreholes 28, 54, 56, 62, 79, 87, 133, 155
bottled water 14, 66–67, 69, 111
   and Flint 17–18, 53, 137, 143, 159
   brands 66, 136–137, 141, 143
   compared with sachets 118
   consumers of 136, 143–44, 147
   cost 66, 69
   environmental impact 137–38
   perceptions 67, 117–118, 137, 147
   production of 138
   safety 67, 143

boutiques 60, 79, 93, 104, 106–07, 111–13, 116, 121, 144

carbon emissions 10
China 36, 46, 49, 101, 120
   automated machines 101, 112, 114, 116
   plastic 36, 54, 64, 78, 82, 128–29
   trade 101, 120, 126
climate 6
climate change 2, 4, 9–11, 27, 55, 77, 138, 156
Commission for Africa 23
commodity chains 98–100, 119, 126
community associations 56, 62–63
conflict 9, 11–12, 42, 78, 88
   over water 2, 12
   crisis 9–10, 138

Flint 4, 18, 137
   water 9–10, 11, 12, 18, 19

deforestation 10
democracy 15, 44, 47, 158, 159
desertification 10, 95

ethnicity 3, 14, 16, 41, 89–93, 94
ethnography 12

finance 28
Flint, Michigan 3–4, 9, 17–18, 45, 137, 143, 159
food security 5, 10, 11, 20
Fulani 11, 16, 41, 75, 79, 90, 91, 93
   gender 90, 145
   identity 80–81, 91, 94
   relations w/Tuareg 87–88, 89
   solidarity 88
   urban context 75, 76, 91, 93–94
   water culture 11, 80, 87–88, 90
   work 59, 71–73, 78–79, 80, 83, 93–94, 111, 154

ga' ruwa 2, 16, 43, 51, 57, 58–60, 64, 66, 69, 71–94, 97, 101, 108, 112, 125–128, 151, 152, 154, 156
   historic roles 43
   gender 3, 11, 89–92, 94, 98, 119, 122
   and Islam 16
   and water 2, 55, 92, 122
   nomadic peoples 74, 81, 92
geography 42, 46–47
global economic order 1
global plastic capitalism 118, 119, 125, 150, 158
Global South 20, 21, 22, 28, 29

good governance 23
governance failure 22, 159
greenhouse gas emissions 10, 137, 138

Hausa
  Ethnicity 34, 41, 49, 51, 53, 59, 71, 72, 74, 80, 81, 83, 90, 92, 93, 94, 104, 105, 108, 110–113, 123, 148
  Language 12, 14, 16, 35, 37, 43, 56, 64, 66, 73, 78, 79, 96, 103, 108, 128, 148–149
  territory (Hausaland) 34, 41, 71
health 1, 2, 4–6, 17, 25, 31, 33, 45, 46, 53, 55, 67, 39, 76, 78, 104, 117, 120, 136, 137, 143, 144, 147, 151, 153–54, 158
hybrid economies 2, 14, 16, 19, 24, 25, 28, 48, 83, 100, 116, 121, 122, 127, 134, 154, 156
hybrid system 14, 24, 25, 26, 28, 77, 101, 121, 122, 125, 150, 156
hydraulic despotism 20
hydraulic landscape 25
hydraulic mission 20
hydrosocial cycle 27, 31
hydrosocial territories 27, 62, 73, 124

inequality 9, 10–11, 12, 14, 19, 46, 55, 154, 155,
informal systems 2, 19, 24, 25, 26, 28, 29, 52, 53, 60, 72, 77, 79, 83, 86–88, 93, 119, 121, 124, 153, 156, 158
Integrated Water Resource Management (IWRM) 19, 23, 24, 157
Intergovernmental Panel on Climate Change (IPCC) 10
International Monetary Fund (IMF) 21, 29, 45, 46, 156
Islam 16, 33, 34, 35, 37, 38, 83, 97, 147, 148
Islamic organizations 36, 37, 44, 54, 63, 133

Jurisdiction 8

Kandadji Dam 12, 157

Mali 11, 41, 42, 72, 74, 75, 77, 78, 79, 82, 87, 88, 90, 91, 92, 93, 157
Mami Wata 34
marabout 36, 80
material 2, 3, 9, 12, 14, 15, 16, 18, 70, 74, 78, 79, 98, 118, 123–150
material culture 2, 16, 18, 31, 32, 55, 65, 98, 123, 125–134, 135–140
materiality 3, 7, 8, 16, 30–32, 38, 123–150, 156, 158, 159
Mawri 41, 42
men 16, 36, 49, 57, 71–74, 78, 79, 81, 87, 90, 92, 94, 97, 107–8, 111
Millenium Development Goals (MDGs) 1, 5, 10, 20, 21
Ministere de l'Hydraulique et d'Assainisement 8, 15
Ministere de l'Hydraulique et de l'Environement 8, 15
mixed methods 12–15
mobile vendors 44, 107–108, 110–112, 121, 126, 138
multinational corporations (MNCs) 9, 18, 24, 28, 29, 44, 46, 155
Muslim 2, 33, 35, 36, 37, 38, 63, 66, 81, 95, 110, 118, 145, 146,

Nestlé 17–18, 159
New Water Decade 20, 21, 22
Niamey Nyala 152–53, 156
Niger River 5, 7, 12, 33, 34, 41, 42, 43, 46, 53, 62, 69, 103, 134, 136, 137, 144, 147, 155, 157, 159
Nigeria 11, 12, 34, 36, 41, 42, 47, 49, 63, 64, 94, 101, 102, 103, 113, 114, 116, 117, 119, 148, 149
nomad 74, 75, 76, 79, 80, 81, 87, 88, 90, 91, 93, 94, 145
non-governmental organization (NGO) 1, 24, 56, 63, 149, 156

participant observation 4
pastoralist 11, 93

Index • 177

plastic vi, 2, 6, 7, 16, 18, 34, 36, 37, 38, 49, 53, 54, 57, 64, 67, 73, 78, 82, 85, 90, 96–150, 152, 158
Pogge, Thomas 1, 154
population growth vii, 6, 20, 29, 40, 43, 45, 46, 49, 77, 85
private sector participation (PSP) 20, 22, 29
privatization 3, 17, 20, 21, 22, 31, 45, 100, 145, 156
pro-poor policies 12, 23, 29, 30, 69, 156, 159
public-private partnerships (PPPs) 9, 19, 21, 23, 121, 122

rain 5, 6, 10, 11, 26, 33, 34, 36, 41, 47, 48, 51, 54, 80, 103, 132

sachets vi, 2, 6, 7, 13, 14, 16, 44, 52, 53, 57, 70, 96–122, 123, 125, 126, 135, 137, 138, 139, 140, 148, 150, 154, 156
Sahara 4, 9, 34, 41, 43, 48, 75, 79, 92, 148
Sahel 4, 6, 9, 34, 41, 47, 48, 51, 68, 74, 79, 80, 89, 123, 133, 134, 150, 156
Sanitation 1, 2, 5, 6, 11, 20, 21, 22, 25, 153
sedentary 11, 74, 75, 79, 88, 90, 91, 92, 93, 94
Société d'Exploitation des Eaux du Niger (SEEN) 8, 15, 45, 56, 62, 68, 69, 72, 83, 85, 129, 145–47

small water enterprises (SWEs) 9, 19
Songhay 33–34, 41, 42, 74, 83, 90, 93, 94, 105, 111
Société de Patrimoine de l'Eau du Niger (SPEN) 8, 15, 45, 67, 153
standpipes 10, 25, 26, 28, 29, 43, 54, 56, 58, 59, 60, 68, 73, 81–85, 97, 105, 125, 129, 136, 153, 155
standpipe managers 58–62, 72, 76, 83–84, 97
Structural Adjustment Programs (SAPs) 15, 21, 28, 45–46, 57

Sustainable Development Goals (SDGs) 1, 145, 159

Tuareg 11, 16, 41, 44, 59, 72, 74–81, 87, 88–94, 154

United Nations Development Program (UNDP) 1
urban Africa 18–19, 21, 26–28, 43, 75, 77, 155–56
urbanization 11, 19, 40, 45, 76

value chains 98–101, 119
values 24, 27, 31, 74, 76, 94, 99, 119
vendors vi, 2, 7, 12–14, 16, 25, 43, 44, 51, 53, 56, 70, 71–95, 96, 101, 102, 106–13, 116, 118–21, 126, 128, 138, 149–150, 152, 156
Veolia 8, 9, 45, 83, 129, 145–46, 158
water
 access 1–7, 9–12, 15, 16, 18, 19, 21, 22, 23, 24, 25–31, 34, 35, 36, 38–39, 41, 43, 51–70, 71, 76–77, 79, 85, 88, 94, 95, 101–102, 122, 123, 124, 125, 133–34, 138, 153–59
 acquisition 12, 27, 28, 31
 availability 1, 5, 9, 10, 19, 25, 28, 31, 70, 80, 133, 144, 157
 commoditization 3, 5–7, 9, 12, 24, 31, 37–39, 55, 74, 77, 94–95, 104, 119, 125, 155, 158, 159
 commodity 3, 5, 7, 8, 21, 31, 44, 94, 98–101, 122, 140, 150
 governance 15, 16, 18–28, 31, 38–39, 119, 121, 150, 153–56, 158
 ground water 5–6
 improved 6, 10–11, 24–25
 myth 33, 37
 policy 5, 15, 18–20, 23–24, 77
 provision 3–4, 10, 11, 18, 20–21, 23–30, 63, 67, 68, 70, 77, 94, 100, 124–25, 155, 158
 purity 16, 53, 98, 104, 117–19, 143–44, 147, 149
 quality 1–3, 5–7, 17–18, 25–30, 77, 103, 104, 133–34, 137–39, 143, 157, 159

water (*cont.*)
- sharing 19, 35–36, 56, 57, 58, 68–69, 156
- subterranean water 6, 9, 133, 155, 157
- surface water 4, 5, 6, 49, 155
- symbol 3, 6, 12, 15–16, 18, 30–39, 74, 76, 95, 98, 106, 118, 119, 126, 136, 140, 144, 145, 153, 154, 157, 158
- tap 4, 17, 54, 56, 57, 59, 62, 64, 66, 67, 69, 70, 88, 89, 117, 124, 129, 133, 143, 151
- value 2, 7, 31, 33, 79, 83, 89, 98–100, 104, 116, 117, 119–20, 122, 134

water access sanitation and health (WASH) 2

waterworlds 156, 159
wells 2, 8, 28, 37, 43, 51, 54, 56, 62, 69, 79, 80, 87–88, 90, 91, 129, 132, 133, 155
women vi, 10, 23, 24, 36, 44, 49, 51, 54, 55, 57, 61, 78, 80, 81, 90–93, 94, 97, 108–10, 113, 118, 154
World Bank 6, 19–21, 23, 29, 30, 44, 45–47, 156
World Health Organization (WHO) 1, 5, 11
World Water Forum 7

Zamzam (Zam-Zam) 37–38, 138, 144–49
Zarma 33, 35, 41–42, 59, 74, 78, 79, 80, 81, 83, 90, 93, 94, 97, 105, 108, 111, 113, 148, 152

Lightning Source UK Ltd.
Milton Keynes UK
UKHW021955181019
351879UK00003B/91/P